An Introduction to Quantum Monte Carlo Methods

An Introduction to Quantum Monte Carlo Methods

Tao Pang
University of Nevada, Las Vegas, USA

Morgan & Claypool Publishers

Rights & Permissions
To obtain permission to re-use copyrighted material from Morgan & Claypool Publishers, please contact info@morganclaypool.com.

ISBN   978-1-6817-4109-3 (ebook)
ISBN   978-1-6817-4045-4 (print)
ISBN   978-1-6817-4237-3 (mobi)

DOI   10.1088/978-1-6817-4109-3

Version: 20161201

IOP Concise Physics
ISSN 2053-2571 (online)
ISSN 2054-7307 (print)

A Morgan & Claypool publication as part of IOP Concise Physics
Published by Morgan & Claypool Publishers, 40 Oak Drive, San Rafael, CA, 94903 USA

IOP Publishing, Temple Circus, Temple Way, Bristol BS1 6HG, UK

To Yunhua, for her love and care

Contents

Preface

In nearly a century, we have witnessed steady progress in the computational study of scientific problems. Now many complex issues in all the technical fields are analyzed and tackled on computers. New paradigms of global-scale computing have emerged, such as the cloud or grid computing. Computers are faster and bigger than ever and come with many more functionalities and applications. There has never been a better time to study scientific problems on computers. Amongst all the computer techniques used in scientific studies, the Monte Carlo approach appears to be most prominent.

This book provides a concise but complete introduction to two computer simulation methods, the diffusion quantum Monte Carlo and the path-integral quantum Monte Carlo, primarily used in research of the many-body problem. There is no assumption of previous experience in computer simulation of the readers but a minimum knowledge of physics typically possessed by an upper-division student or a beginning graduate in physics is required.

To make this book practical, two complete programs in Java, one for the diffusion quantum Monte Carlo simulation of ^{4}He clusters on a graphite surface and the other for the path-integral quantum Monte Carlo simulation of cold atoms in a potential trap, are ready to be downloaded and altered for any research project that the reader wants. These programs will be maintained and improved over time. There will also be additions to the existing programs and they are all accessible through my web page: http://www.physics.unlv.edu/~pang.

Tao Pang

Las Vegas, Nevada, USA

August, 2016

Acknowledgements

Most materials presented in this book have been influenced strongly by my research in the last thirty years. I am extremely grateful to the University of Minnesota; Miller Institute for Basic Research in Science at the University of California, Berkeley; National Science Foundation; Department of Energy; and W.M. Keck Foundation for their generous support of my research over those years.

Many colleagues from all over the world have made contributions to this book either through collaborating with me on some of the projects described here or communicating with me on the subjects discussed in this book. My deepest gratitude goes to those who have worked or communicated with me over the years regarding the topics covered in the book, especially to those inspired young scholars who have constantly reminded me that the effort of writing this book will be worthwhile and the students who have taken courses from me on related matters.

Author biography

Tao Pang

Tao Pang is Professor of Physics at the University of Nevada, Las Vegas (UNLV). Following his higher education at Fudan University, one of the most prestigious institutions in China, he obtained his PhD in condensed matter theory from the University of Minnesota in 1989. He then spent two years as a Miller Research Fellow at the University of California, Berkeley, before joining the UNLV physics faculty in the fall of 1991. He has been Professor of Physics at UNLV since 2002. His main areas of research include condensed matter theory and computational physics.

IOP Concise Physics

An Introduction to Quantum Monte Carlo Methods

Tao Pang

Chapter 1

Introduction

One of the *grand challenges* in physics is the many-body problem. Its solution requires a systematic approach to strongly interacting quantum systems with accuracy up to any desired level. Over nearly a hundred years after the discovery of quantum principles, great progress has been made in understanding phenomena associated with quantum particles such as structures of nuclei, atoms, molecules, and crystals. However, our knowledge of strongly interacting quantum systems is still quite limited, even though certain many-body phenomena, such as the conventional superconductivity, metal–insulator transition, and fractional quantum Hall effect, are explained well with different theoretical tools to a satisfactory level.

Understanding many-body quantum systems usually means solving the Schrödinger equation of a system of many strongly interacting particles. However, there is no panacea for solving such an equation. Different approximations must be made in order to understand the roles played by the strong interactions between the particles in different systems under various conditions. Thus numerical approaches become prominent in this area. Two sampling schemes based on the Metropolis algorithm are extremely noticeable, that is, the diffusion and path-integral quantum Monte Carlo methods.

The diffusion quantum Monte Carlo method is a stochastic scheme (Anderson 2007) that can extract important information about the ground state, and sometimes excited states, of a many-body system. The method maps the Schrödinger equation to a diffusion equation under an imaginary-time transformation and guides the system toward its ground state over the evolution of time. In principle, it is an exact method for calculating the ground-state properties of many-boson systems.

The path-integral quantum Monte Carlo method (Kashiwa *et al* 1997) is based on the Feynman path-integral formulation of quantum statistics of a many-body system at finite temperature. The method decomposes the density matrix of a many-body system into a path integral and is extremely suitable to studying the temperature-dependent properties of many-body systems. The real-time path-integral approach

doi:10.1088/978-1-6817-4109-3ch1 1-1

also shows promise in studying the dynamics of many-body systems (Mühlbacher and Rabani 2007).

In both the methods, there is a major hurdle that forbids us from applying them to many-fermion systems to a desired accuracy because of the Fermi statistics. A typical wavefunction or density matrix of a many-fermion system has a complex nodal structure that prevents an easy stochastic process from driving the system to its ground state or desired excited state. This problem, originating from the Pauli exclusion principle of fermions, is known as the *fermion-sign problem*, because crossing a node of the wavefunction or density matrix changes its sign, and thus disallows a straightforward weight interpretation of the wavefunction or density matrix. It has been shown that this fermion-sign problem has the NP complexity by Troyer and Wiese (2005) and the NP problem at the moment is unsolved mathematically.

1.1 Sampling

Before we go into the actual simulation with the Metropolis algorithm, let us first get an idea of the concept of the Monte Carlo method of quadrature for evaluating multi-variable integrals. Here we use a simple example of a two-variable integral to illustrate how a basic Monte Carlo scheme works. If we want to evaluate the two-dimensional integral

$$S = \int_0^1 \int_0^1 f(x, y)\mathrm{d}x\, \mathrm{d}y, \tag{1.1}$$

we can simply divide the integration region of the square evenly into M_x slices along the x direction and M_y slices along the y direction, and the integral can then be approximated as

$$S = \frac{1}{M} \sum_{j=1}^{M_y} \sum_{i=1}^{M_x} f\left(x_i,\, y_j\right) + O(h^2) \tag{1.2}$$

under the trapezoid rule. This sum is equivalent to sampling the integration region from points (x_1, y_1), (x_2, y_1), ..., (x_{M_x}, y_{M_y}) with an equal weight, in this case, $1/M$, at each point. Note that $M = M_x M_y$ and h is on the order of $1/M_x$ or $1/M_y$, whichever is greater.

We can, on the other hand, select M sets of x_i and y_j from a uniform random-number generator from 0 to 1 to accomplish the same goal. If M is very large, we would expect the selected x_i and y_j to be uniformly distributed in region [0, 1] with fluctuations proportional to $1/\sqrt{M}$. Then the integral can be approximated by the average

$$S \simeq \frac{1}{M} \sum_{i=1}^{M} f(x_i, y_i), \tag{1.3}$$

where x_i and y_i are M sets of points generated uniformly on the square. Note that the possible error in the evaluation of the integral is now given by the fluctuation of the distribution (x_i, y_i). If we use the standard deviation of statistics to estimate the possible error of the random sampling, we have

$$(\Delta S)^2 = \frac{1}{M} \left(\langle f \rangle^2 - \langle f \rangle^2 \right). \tag{1.4}$$

Here the average of a quantity is defined as

$$\langle A \rangle = \frac{1}{M} \sum_{i=1}^{M} A_i, \tag{1.5}$$

with A_i being the sampled data.

Now we would like to illustrate how the scheme works in an actual numerical example. In order to demonstrate the algorithm clearly, let us take a very simple integrand $f(x, y) = x^2(1 - y^2)$. The exact result of the integral is $(1 - 1/3)/3 = 0.222...$ for this simple example. The following program implements the basic sampling scheme discussed above.

```
// An example of integration with direct Monte Carlo scheme
// with integrand f(x) = x*x*(1-y*y).

import java.lang.*;
import java.util.Random;
public class Monte {
  public static void main(String argv[]) {
    Random r = new Random();
    int M = 1000000;
    double s0 = 0;
    double ds = 0;
    for (int i=0; i<M; ++i) {

        double x = r.nextDouble();
        double y = r.nextDouble();
        double f = x*x*(1-y*y);
        s0 += f;
        ds += f*f;
    }
    s0 /= M;
    ds /= M;
    ds = Math.sqrt(Math.abs(ds-s0*s0)/M);
    System.out.println("S = " + s0 + " +- " + ds);
   }
  }
```

We have used the uniform random-number generator in Java for convenience. The numerical result with its estimated error from the above program is 0.2220 ± 0.0002. More reliable results can be obtained from the average of several independent runs. For example, with four independent runs, we have obtained an average of 0.2222 with an estimated error reduced to 0.0001.

The simple Monte Carlo quadrature used in the above program does not show any advantage. For example, the trapezoid rule yields a higher accuracy with the same number of points. The reason is that the error from the Monte Carlo quadrature is

$$\Delta S \propto \frac{1}{M^{1/2}}, \tag{1.6}$$

whereas the trapezoid rule yields an estimated error of

$$\Delta S \propto \frac{1}{M}, \tag{1.7}$$

in two dimensions, which is much smaller for a large M. The true advantage of the Monte Carlo method is in the evaluation of multidimensional integrals of a large number of variables. For example, if we are interested in a many-body system, such as the electrons in a neon atom, we have 10 particles, and the integral for the expectation value can be as high as 30 dimensions. For a d-dimensional space, the Monte Carlo quadrature will still yield the same error behavior, that is, proportional to $1/\sqrt{M}$, with M being the number of points sampled. However, the corresponding error in the trapezoid rule becomes

$$\Delta S \propto \frac{1}{M^{2/d}}, \tag{1.8}$$

which is greater than the Monte Carlo error estimate with the same number of points if d is greater than four. The point is that the Monte Carlo quadrature produces a much more reasonable estimate of a d-dimensional integral when d is very large. There are some specially designed numerical quadratures that would still work better than the Monte Carlo quadrature when d is slightly larger than four. But for a real many-body system, the dimensionality in an integral is $3N$, where N is the number of particles in the system. When N is on the order of 10 or larger, any other workable quadrature would perform worse than the Monte Carlo quadrature.

1.2 Random-number generators

One important ingredient in the above sampling process is to have a high-quality uniform random-number generator. Similar to the example discussed above, many numerical simulations will need random-number generators of different distributions either in setting up the initial configuration or for creating new configurations of the system. However, there is no such thing as *random* in a computer program. A computer will execute the program exactly the same if the program is started the same. A random-number generator here really means a pseudo-random-number generator that can generate a long sequence of numbers that can imitate a given distribution. In this section we will discuss some basic random-number generators used in computational physics and other computer simulations.

Uniform random-number generators

The most useful random-number generators are those with a uniform distribution in a given region. The three most important criteria for a good uniform random-number generator are the following (Park and Miller 1988).

First, a good generator should have a *long period*, which should be close to the range of the integers adopted. For example, if 32-bit integers are used, a good generator should have a period close to $2^{31} - 1 = 2\,147\,483\,647$. The range of the 32-bit integers is $[-2^{31}, 2^{31} - 1]$. Note that one bit is used for the signs of the integers.

Second, a good generator should have the best *randomness*. There should only be a very small correlation among all the numbers generated in sequence. If $\langle A \rangle$ represents the statistical average of the variable A, the k-point correlation function of the numbers generated in the sequence $\langle x_{i_1} x_{i_2} \cdots x_{i_k} \rangle$ for $k = 2, 3, 4, \ldots$, should be very small. One way to illustrate the behavior of the correlation function $\langle x_i x_{i+l} \rangle$ is to plot x_i and x_{i+l} in an xy plane. A good random-number generator will have a very uniform distribution of the points for any $l \neq 0$. A poor generator may show stripes, lattices, or other inhomogeneous distributions.

Finally, a good generator has to be very *fast*. In practice, we will need a lot of random numbers in order to have good statistical results. The speed of the generator can become a very important factor.

The simplest uniform random-number generator is built from the so-called linear congruent scheme. The random numbers are generated in sequence from the linear relation

$$x_{i+1} = (a\,x_i + b) \bmod c, \tag{1.9}$$

where a, b, and c are *magic numbers*: their values determine the quality of the generator. One common choice, $a = 7^5 = 16\,807$, $b = 0$, and $c = 2^{31} - 1 = 2\,147\,483\,647$, has been tested and found to be excellent for generating unsigned 32-bit random integers. It has the full period of $2^{31} - 1$ and is very fast. The correlation function $\langle x_{i_1} x_{i_2} \ldots x_{i_k} \rangle$ is very small. In figure 1.1, we plot x_i and x_{i+10} (normalized by c) generated using the linear congruent method with the above selection of the magic numbers. Note that the plot is very homogeneous and random. There are no stripes, lattice structures, or any other visible patterns in the plot.

Implementation of this random-number generator on a computer is not always trivial, because of the different numerical range of the integers specified by the computer language or hardware. For example, most 32-bit computers would have integers in $[-2^{31}, 2^{31} - 1]$. If a number runs out of this range by accident, the computer will reset it to zero. If the computer could have modulated the integers by $2^{31} - 1$ automatically, we would have implemented a random-number generator with the above magic numbers a, b, and c simply by taking consecutive numbers

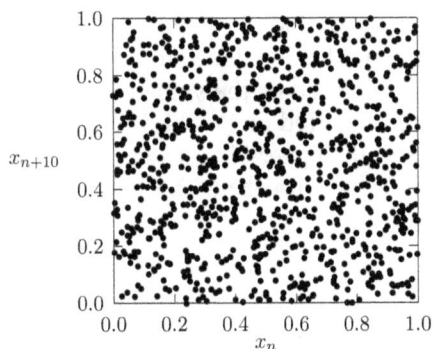

Figure 1.1. 1000 points of the random-number pairs (x_i, x_{i+10}) normalized to the range of $[0, 1]$ are shown.

from $x_{i+1} = 16\,807\,x_i$ with any initial choice of $1 < x_0 < 2^{31} - 1$. The range of this generator would be $[0, 2^{31} - 1]$. However, this automatic modulation would cause some serious problems in other situations. For example, when a quantity is out of range due to a bug in the program, the computer would still wrap it back without producing any error messages. This is why, in practice, computers do not modulate numbers automatically, so we have to devise a scheme to modulate the numbers generated in sequence. The following method is an implementation of the uniform random-number generator (normalized to the range of $[0, 1]$) discussed above.

```
// Method to generate a uniform random number in [0,1]
// following x(i+1)=a*x(i) mod c with a=pow(7,5) and
// c=pow(2,31)-1.  Here the seed is a global variable.

public static double ranf() {
    final int a = 16807, c = 2147483647, q = 127773,
       r = 2836;
    final double cd = c;
    int h = seed/q;
    int l = seed%q;
    int t = a*l - r*h;
    if (t > 0) seed = t;
    else seed = c + t;
    return seed/cd;
}
```

Note that in the above code, we have used two more magic numbers $q = c/a$ and $r = c \bmod a$. A program in Pascal similar to the above method is given by Park and Miller (1988). We can easily show that the above method would modulate the numbers with $c = 2^{31} - 1$ on any computer with integers of 32 bits or more. To use this method, the seed needs to be a global variable that is returned each time when the method is called. To show that the method given here does implement the algorithm correctly, we can set the initial seed to be 1, and then after 10 000 steps, we should have 1 043 618 065 returned as the value of the seed.

The above generator has a period of $2^{31} - 1$. If a longer period is desired, we can create similar generators with higher-bit integers. For example, we can create a generator with a period of $2^{63} - 1$ for the 64-bit integers with the following method.

```
// Method to generate a uniform random number in [0,1]
// following x(i+1)=a*x(i) mod c with a=pow(7,5) and
// c=pow(2,63)-1.  Here the seed is a global variable.

public static double ranl() {
    final long a = 16807L, c = 9223372036854775807L,
        q = 548781581296767L, r = 12838L;
    final double cd = c;
    long h = seed/q;
    long l = seed%q;
    long t = a*l - r*h;
    if (t > 0) seed = t;
    else seed = c + t;
    return seed/cd;
}
```

Note that we have used $a = 7^5$, $c = 2^{63} - 1$, $q = c/a$, and $r = c \bmod a$ in the above method with the seed being a 64-bit (long) integer.

In order to start the random-number generator differently every time, we need to have a systematic way of obtaining a different initial seed. Otherwise, we would not be able to obtain fair statistics. Almost every computer language has intrinsic routines to report the current time in an integer form, and we can use this integer to construct an initial seed (Anderson 1990).

For example, most computers can produce $0 \leqslant t_1 \leqslant 59$ for the second of the minute, $0 \leqslant t_2 \leqslant 59$ for the minute of the hour, $0 \leqslant t_3 \leqslant 23$ for the hour of the day, $1 \leqslant t_4 \leqslant 31$ for the day of the month, $1 \leqslant t_5 \leqslant 12$ for the month of the year, and t_6 for the current year in common era. Then we can choose

$$i_s = t_6 + 70\big(t_5 + 12\{t_4 + 31[t_3 + 23(t_2 + 59t_1)]\}\big) \qquad (1.10)$$

as the initial seed, which is roughly in the region of $[0, 2^{31} - 1]$. The results should never be the same as long as the jobs are started at least a second apart.

We would like to demonstrate how to apply the random-number generator here and how to initiate the generator with the current time through a simple example. Consider evaluating π by randomly throwing a dart to a unit square defined by $x \in [0, 1]$ and $y \in [0, 1]$. The chance for the dart to land inside the quarter of the unit circle centered at the origin of the coordinates is $\pi/4$ after comparing the areas of the unit square and the quarter of the unit circle. The following program is an implementation of such an evaluation of π in Java.

```java
// An example of evaluating pi by throwing a dart to a
// unit square with 0<x<1 and 0<y<1.

import java.lang.*;
import java.util.Calendar;
import java.util.GregorianCalendar;
public class Dart {
  final static int n = 1000000;
  static int seed;
  public static void main(String argv[]) {

// Initiate the seed from the current time
    GregorianCalendar t = new GregorianCalendar();
    int t1 = t.get(Calendar.SECOND);
    int t2 = t.get(Calendar.MINUTE);
    int t3 = t.get(Calendar.HOUR_OF_DAY);
    int t4 = t.get(Calendar.DAY_OF_MONTH);
    int t5 = t.get(Calendar.MONTH)+1;
    int t6 = t.get(Calendar.YEAR);
    seed = t6+70*(t5+12*(t4+31*(t3+23*(t2+59*t1))));
    if ((seed%2) == 0) seed = seed-1;

// Throw the dart to the unit square
    int ic = 0;
    for (int i=0; i<n; ++i) {
      double x = ranf();
      double y = ranf();
      if ((x*x+y*y) < 1) ic++;
    }
    System.out.println("Estimated pi: " + (4.0*ic/n));
  }

  public static double ranf() {...}

}
```

An even initial seed is usually avoided in order to have the full period of the generator realized. Note that the month in Java is recorded between 0 and 11; so we add it by 1 in the above program. To initiate the 64-bit generator, we can use the method `getTime()` from the `Date` class in Java, which returns the current time in milliseconds in a 64-bit (`long`) integer, measured from midnight January 1, 1970.

Uniform random-number generators are very important in scientific computing, and good ones are extremely difficult to find. The generator given here is considered to be one of the best uniform random-number generators.

New computer programming languages such as Java typically come with a comprehensive, intrinsic set of random-number generators that can be initiated automatically with the current time or with a chosen initial seed. We will demonstrate the use of such a generator in Java later in this section.

Other distributions

As soon as we obtain good uniform random-number generators, we can use them to create other types of random-number generators. For example, we can use a uniform random-number generator to create an exponential distribution or a Gaussian distribution.

All the exponential distributions can be cast into their simplest form

$$p(x) = e^{-x} \qquad (1.11)$$

after proper choice of units and coordinates. For example, if a system has energy levels of E_0, E_1, ..., E_n, the probability for the system to be at the energy level E_i at temperature T is given by

$$p(E_i, T) \propto e^{-(E_i - E_0)/kT}, \qquad (1.12)$$

where k is the Boltzmann constant. If we choose kT as the energy unit and E_0 as the zero point, the above equation reduces to equation (1.11).

One way to generate the exponential distribution is to relate it to a uniform distribution. For example, if we have a uniform distribution $f(y) = 1$ for $y \in [0, 1]$, we can relate it to an exponential distribution by

$$f(y)dy = dy = p(x)dx = e^{-x}dx, \qquad (1.13)$$

which gives

$$y(x) - y(0) = 1 - e^{-x} \qquad (1.14)$$

after integration. We can set $y(0) = 0$ and invert the above equation to have

$$x = -\ln(1 - y), \qquad (1.15)$$

which relates the exponential distribution of $x \in [0, \infty]$ to the uniform distribution of $y \in [0, 1]$. The following method is an implementation of the exponential random-number generator given in the above equation and constructed from a uniform random-number generator.

```
// Method to generate an exponential random number from a
// uniform random number in [0,1].

    public static double rane() {
        return -Math.log(1-ranf());
    }

    public static double ranf() {...}
```

The uniform random-number generator obtained earlier is used by this method. Note that when this method is used in a program, the seed still has to be a global variable.

As we have pointed out, another useful distribution used in physics is the Gaussian distribution

$$g(x) = \frac{1}{\sqrt{2\pi\sigma}}e^{-x^2/2\sigma^2}, \tag{1.16}$$

where σ is the variance of the distribution, which we can take as 1 for the moment. The distribution with $\sigma \neq 1$ can be obtained via the rescaling of x by σ. We can use a uniform distribution $f(\phi) = 1$ for $\phi \in [0, 2\pi]$ and an exponential distribution $p(t) = e^{-t}$ for $t \in [0, \infty]$ to obtain two Gaussian distributions $g(x)$ and $g(y)$. We can relate the product of a uniform distribution and an exponential distribution to a product of two Gaussian distributions by

$$\frac{1}{2\pi}f(\phi)d\phi\,p(t)dt = g(x)dx\,g(y)dy, \tag{1.17}$$

which gives

$$e^{-t}dt\,d\phi = e^{-(x^2+y^2)/2}dx\,dy. \tag{1.18}$$

The above equation can be viewed as the coordinate transform from the polar system (ρ, ϕ) with $\rho = \sqrt{2t}$ into the rectangular system (x, y), that is,

$$x = \sqrt{2t}\cos\phi, \tag{1.19}$$

$$y = \sqrt{2t}\sin\phi, \tag{1.20}$$

which are two Gaussian distributions if t is taken from an exponential distribution and ϕ is taken from a uniform distribution in the region $[0, 2\pi]$. With the availability of the exponential random-number generator and uniform random-number generator, we can construct two Gaussian random numbers immediately from equations (1.19) and (1.20). The exponential random-number generator itself can be obtained from a uniform random-number generator as discussed above. Here we would like to show how to create two Gaussian random numbers from two uniform random numbers.

```
// Method to create two Gaussian random numbers from two
// uniform random numbers in [0,1].

  public static double[] rang() {
     double x[] = new double[2];
     double r1 = - Math.log(1-ranf());
     double r2 = 2*Math.PI*ranf();
     r1 = Math.sqrt(2*r1);
     x[0] = r1*Math.cos(r2);
     x[1] = r1*Math.sin(r2);
     return x;
  }


  public static double ranf() {...}
```

In principle, we can generate any given distribution numerically. For the cases of the Gaussian distribution and exponential distribution, we construct the new generators with the integral transformations in order to relate the distributions sought to the distributions known. A general procedure can be devised by dealing with the integral transformation numerically. For example, we can use the Metropolis algorithm, to be discussed in chapter 2, to obtain any distribution numerically.

Percolation in two dimensions

Let us use two-dimensional percolation as an example to illustrate how a random-number generator is utilized in computer simulations. When atoms are added to a solid surface, they will first occupy the sites with the lowest potential energy to form small two-dimensional clusters. If the probability of occupying each empty site is still high, the clusters will grow on the surface and eventually form a single layer, or a percolated two-dimensional network. If the probability of occupying each empty site is low, the clusters will grow into island-like three-dimensional clusters. The general problem of the formation of a two-dimensional network can be cast into a simple model with each site carrying a fixed occupancy probability.

Assume that we have a two-dimensional square lattice with $n \times n$ lattice points. Then we can generate $n \times n$ random numbers $x_{ij} \in [0, 1]$ for $i = 0, 1, ..., n - 1$ and $j = 0, 1, ..., n - 1$. The random number x_{ij} is further compared with the assigned occupancy probability $p \in [0, 1]$. The site is occupied if $p > x_{ij}$; otherwise the site remains empty. Clusters are formed by the occupied sites. A site in each cluster is, at least, a nearest neighbor of another site in the same cluster. We can gradually change p from 0 to 1. As p increases, the sizes of the clusters will increase and some clusters will also merge into larger clusters. When p reaches a critical probability p_c, there will be one cluster of the occupied sites, which will reach all the boundaries of the lattice. We call p_c the percolation threshold. The following method is the core part of the simulation of two-dimensional percolation, which assigns a false value to an empty site and a true value to an occupied site.

```
// Method to create a 2-dimensional percolation lattice.

import java.util.Random;
  public static boolean[][] lattice(double p, int n) {
    Random r = new Random();
    boolean y[][] = new boolean[n][n];
    for (int i=0; i<n; ++i) {
      for (int j=0; j<n; ++j) {
        if (p > r.nextDouble()) y[i][j] = true;
        else y[i][j] = false;
      }
    }
    return y;
  }
```

Here y_{ij} is a Boolean array that contains true values at all the occupied sites and false values at all the empty sites. We can use the above method with a program that has p increased from 0 to 1, and can sort out the sizes of all the clusters formed by the occupied lattice sites. In order to obtain good statistical averages, the procedure should be carried out many times. For more discussions on percolation, see Stauffer and Aharony (1992) and Grimmett (1999). Note that we have used the intrinsic random-number generator from Java. nextDouble() is a method in the Random class, and it creates a floating-point random number in [0, 1] when it is called. The default initiation of the generator, as used above, is from the current time.

The above example is an extremely simple application of the uniform random-number generator in physics. In the following chapters, we will discuss the application of random-number generators in other simulations. Interested readers can find more on different random-number generators in Knuth (1998), Park and Miller (1988), and Anderson (1990).

Exercises

1.1 Apply the Monte Carlo quadrature to evaluate integral

$$ S = \int_0^1 \int_0^1 x^2 e^{-(x^2+y^2)} \mathrm{d}x\, \mathrm{d}y. $$

1.2 Evaluate the integral

$$S = \int (xy + yz)r^2 e^{-r^2} dx \, dy \, dz$$

with the Monte Carlo sampling in the entire three-dimensional space with $r^2 = x^2 + y^2 + z^2$. One way to deal with the infinite space is to truncate to a finite space and then extrapolate the evaluation to the infinite space.

1.3 One way to calculate π is by randomly throwing a dart to the unit cube defined by $x \in [0, 1]$, $y \in [0, 1]$, and $z \in [0, 1]$, in three-dimensional space. The chance for the dart to land inside the unit sphere centered at the origin of the coordinates is $\pi/6$, from comparison of the volume of the eighth of the sphere and the unit cube. Write a program to calculate π in such a manner.

1.4 If we randomly drop a needle of unit length into an infinite plane that is covered by parallel lines one unit apart, the probability for the needle to land in a gap (not crossing any line) is $1 - 2/\pi$. Derive this probability analytically. Write a program that can sample the needle dropping process and calculate the probability without explicitly using the value of π. Compare your numerical result with the analytical result and discuss possible sources of error.

1.5 Generate 21 pairs of random numbers (x_i, f_i) in $[0, 1]$ and sort them according to $x_{i+1} \geqslant x_i$. Treat them as a discrete set of function $f(x)$ and fit them to $p_{20}(x)$, where

$$p_m(x) = \sum_{k=0}^{m} \alpha_k u_k(x),$$

with $u_k(x)$ being orthogonal polynomials.

1.6 Develop a scheme that can generate any distribution $w(x) > 0$ in a given region $[a, b]$. Implement the scheme in a method and test it with $w(x) = 1$, $w(x) = e^{-x^2}$, and $w(x) = x^2 e^{-x^2}$. Vary a and b, and compare the results here with the uniform random-number generator and the Gaussian random-number generator given in this chapter.

1.7 Generate a large set of Gaussian random numbers and sort them into an increasing order. Then count the data points falling into each of the uniformly divided intervals. Use these values to perform a least-squares fit of the generated data to the function $f(x) = ae^{-x^2/2\sigma^2}$, where a and σ are parameters to be determined. Comment on the quality of the generator based on the fitting result.

1.8 Write a program that can generate clusters of occupied sites in a two-dimensional square lattice with $n \times n$ sites. Determine $p_c(n)$, the threshold probability with at least one cluster across the whole lattice. Then determine p_c for an infinite lattice from

$$p_c(n) = p_c + \frac{a_1}{n} + \frac{a_2}{n^2} + \frac{a_3}{n^3} + \cdots,$$

where a_i and p_c can be solved from the $p_c(n)$ obtained.

Bibliography

Anderson J C 2007 *Quantum Monte Carlo: Origins, Development, Applications* (New York: Oxford University Press)

Anderson S L 1990 Random number generators on vector supercomputers and other advanced architectures *SIAM Rev.* **32** 221–51

Grimmett G 1999 *Percolation* (Berlin: Springer)

Kashiwa T, Ohnuki Y and Suzuki M 1997 *Path Integral Methods* (New York: Oxford University Press)

Knuth D E 1998 *The Art of Computer Programming Seminumerical Algorithms* vol 2 (Reading, MA: Addison-Wesley)

Mühlbacher L and Rabani E 2007 Real-time path integral approach to nonequilibrium many-body quantum system *Phys. Rev. Lett.* **100** 176403

Park S K and Miller K W 1988 Random number generators: good ones are hard to find *Commun. ACM* **31** 1192–201

Stauffer D and Aharony A 1992 *Introduction to Percolation Theory* (London: Taylor and Francis)

Troyer M and Wiese U-J 2005 Computational complexity and fundamental limitations to fermionic quantum Monte Carlo simulations *Phys. Rev. Lett.* **94** 170201

An Introduction to Quantum Monte Carlo Methods

Tao Pang

Chapter 2

The Metropolis algorithm

Here we would like to discuss a sampling scheme devised by Metropolis *et al* (1953) in order to evaluate multivariable integrals under a distribution function more effectively than a primitive Monte Carlo quadrature. The method takes great advantage of the fast-changing distribution function; a typical problem in statistical mechanics that requires carrying out the evaluation of a physical quantity by integration over many variables under a sharp Boltzmann distribution. The essence of the Metropolis algorithm is to use importance sampling; namely, selecting the points in the configuration space according to the distribution function of the physical system under study.

2.1 Importance sampling

Consider a system of N particles, such as the atoms or molecules in a classical liquid, with $\mathbf{R} = (\mathbf{r}_1, \mathbf{r}_2, \ldots, \mathbf{r}_N)$ being the $3N$-dimensional position vector of the particles in the system. In the canonical ensemble, we deal with a statistical process that leads the system to the equilibrium distribution of the particles as

$$W(\mathbf{R}) = \frac{e^{-U(\mathbf{R})/kT}}{\int e^{-U(\mathbf{R}')/kT} d\mathbf{R}'}, \tag{2.1}$$

where $U(\mathbf{R})$ is the potential energy of the system for the given configuration $\mathbf{R}$. Here k is the Boltzmann constant and T is the temperature of the system. We have ignored the momentum or velocity dependence of the distribution function, with the understanding that such a dependence is dealt with analytically. We also assume a normalized distribution; that is,

$$\int W(\mathbf{R}) d\mathbf{R} = 1, \tag{2.2}$$

doi:10.1088/978-1-6817-4109-3ch2

for our convenience. The average of a physical quantity A of the system is given by

$$\langle A \rangle = \int A(\mathbf{R})W(\mathbf{R})\, d\mathbf{R} \simeq \frac{1}{M}\sum_{i=1}^{M} A(\mathbf{R}_i), \qquad (2.3)$$

where M is the total number of configurations of $\mathbf{R}_i$, for $i = 1, 2, ..., M$, sampled according to the distribution function $W(\mathbf{R}_i)$. How to sample this distribution function most effectively is the motivation of the Metropolis algorithm.

In equilibrium, the values of the distribution function at different points of the configuration space must satisfy the condition

$$W(\mathbf{R})T(\mathbf{R} \rightarrow \mathbf{R}') = W(\mathbf{R}')T(\mathbf{R}' \rightarrow \mathbf{R}), \qquad (2.4)$$

where $T(\mathbf{R} \rightarrow \mathbf{R}')$ is the transition rate from configuration $\mathbf{R}$ to configuration $\mathbf{R}'$. This condition is usually referred to as *detailed balance*.

The *Metropolis algorithm* samples the configurations such that the move from one configuration $\mathbf{R}$ to another $\mathbf{R}'$ is accepted if the ratio of the transition rates satisfies the condition

$$\frac{T(\mathbf{R} \rightarrow \mathbf{R}')}{T(\mathbf{R}' \rightarrow \mathbf{R})} = \frac{W(\mathbf{R}')}{W(\mathbf{R})} > \eta, \qquad (2.5)$$

where $\eta \in [0, 1]$ is a uniform random number. To evaluate the integral $\langle A \rangle$, we first randomly select a configuration $\mathbf{R}$ that is allowable and, if possible, close to having the highest probability, and then evaluate $W(\mathbf{R})$. A new configuration $\mathbf{R}'$ is attempted with

$$\mathbf{R}' = \mathbf{R} + \Delta\mathbf{R}, \qquad (2.6)$$

where $\Delta\mathbf{R}$ is a $3N$-dimensional vector with a component chosen from a uniform distribution between $[-h, h]$. For example,

$$\Delta x_i = h(2\eta - 1), \qquad (2.7)$$

for the x component of $\mathbf{r}_i$. If the length scale in one direction, for example, the z direction, is different from the others, we can choose h_z to be different from h_x or h_y. The value chosen for the step size h is determined from the desired acceptance probability (the ratio of the accepted to the attempted steps). Larger h results in a smaller acceptance probability. To optimize the sampling, the acceptance probability times h^2 is maximized. In practice, h is commonly chosen so that the acceptance probability of the moves is less than or close to 50%. The attempted change in the configuration is usually made by moving one particle at a time. However, moving all particles at once can speed up the exploration of configuration space if N is not too large. In the program provided all the components of $\mathbf{R} = (\mathbf{r}_1, \mathbf{r}_2, ..., \mathbf{r}_N)$ are attempted at each Metropolis move.

The ratio of the distribution function at the attempted new configuration to that of the old configuration is computed and the new configuration is accepted using equation (2.5). If the ratio is greater than η, the new configuration is accepted; otherwise, the old configuration is assumed to be the new configuration. The desired quantity $A(\mathbf{R}_i)$ is evaluated at $\ell = n_1 + n_0, n_1 + 2n_0, \ldots, n_1 + Mn_0$, and the integral is approximated by

$$\langle A \rangle \simeq \frac{1}{M} \sum_{\ell=1}^{M} A(\mathbf{R}_{n_1+\ell n_0}). \tag{2.8}$$

Note that $n_1 + n_0$ moves are used to remove the influence of the initial configuration. The thermodynamic data are taken n_0 moves apart to avoid correlations between the data because they are generated consecutively and a reasonable number of attempted moves, typically on the order of 10 (Pang 2006), should be skipped before the next data point is used.

In most cases, the distribution function $W(\mathbf{R})$ varies by several orders of magnitude, whereas $A(\mathbf{R})$ remains smooth or is nearly constant. The sampling scheme we have outlined yields the average of a physical quantity according to the distribution $W(\mathbf{R})$. In other words, configurations with a greater value of $W(\mathbf{R})$ will occur more often. Importance sampling is the essence of the Metropolis algorithm and is much more efficient than a random sampling of an integrand with a large number of variables.

Now we would like to compare this procedure numerically with direct, random sampling. We still consider the integral

$$S = \int_0^1 \int_0^1 f(x, y) \, dx \, dy = \int_0^1 \int_0^1 W(x, y) g(x, y) \, dx \, dy, \tag{2.9}$$

with $f(x, y) = x^2(1 - y^2)$. We can choose the distribution function as

$$W(x, y) = \frac{1}{Z}\left(e^{x^2} - 1\right), \tag{2.10}$$

which is positive definite. The normalization constant Z is given by

$$Z = \int_0^1 \int_0^1 \left(e^{x^2} - 1\right) dx \, dy = 0.46265167, \tag{2.11}$$

which is calculated from another numerical scheme for convenience. Then the corresponding function $g(x, y) = f(x, y)/W(x, y)$ is given by

$$g(x, y) = Z\frac{x^2(1 - y^2)}{e^{x^2} - 1}. \tag{2.12}$$

Now we are ready to put all of these into a program that is a realization of the Metropolis algorithm for the integral specified.

```java
// An example of Monte Carlo simulation with the
// Metropolis scheme with integrand f(x,y) = x*x*(1-y*y).

import java.lang.*;
import java.util.Random;
public class Carlo {
  static final int nsize=10000;
  static final int nskip=15;
  static final int ntotal=nsize*nskip;
  static final int neq=10000;
  static int iaccept=0;
  static double x, y, w, h=0.4, z=0.46265167;
  static Random r=new Random();
  public static void main(String argv[]) {
    x = r.nextDouble();
    y = r.nextDouble();
    w = weight();
    for (int i=0; i<neq; ++i) metropolis();

    double s0 = 0;
    double ds = 0;
    iaccept = 0;
    for (int i=0; i<ntotal; ++i) {
      metropolis();
      if (i%nskip==0) {
        double f = g(x,y);
        s0 += f;

        ds += f*f;
      }
    }
    s0 /= nsize;
    ds /= nsize;
    ds = Math.sqrt(Math.abs(ds-s0*s0)/nsize);
    s0 *= z;
    ds *= z;
    double accept = 100.0*iaccept/(ntotal);
    System.out.println("S = " + s0 + " +- " + ds);
    System.out.println("Accept rate = " + accept + "%");
  }
```

```
public static void metropolis() {
  double xold=x, yold=y;
  x += 2*h*(r.nextDouble()-0.5);
  y += 2*h*(r.nextDouble()-0.5);
  if (((x<0)||(x>1))||((y<0)||(y>1))) {
    x = xold;
    y = yold;
  }
  else {
    double wnew=weight();
    if (wnew>w*r.nextDouble()) {
      w = wnew;
     ++iaccept;
    }
    else {
      x=xold;
      y=yold;
    }
  }
}

public static double weight() {
  return Math.exp(x*x)-1;
}

public static double g(double xv, double yv) {
  return xv*xv*(1-yv*yv)/(Math.exp(xv*xv)-1);

  }
}
```

The numerical result obtained with the above program is 0.222 ± 0.001. The step size is adjustable, and we should try to keep it such that the accepting rate of the new configurations is less than or around 50%. In practice, it seems that such a choice of accepting rate is compatible with considerations of speed and accuracy. The above program is structured so that we can easily modify it to study other problems. The result here does not appear to be impressive but the strength of the Metropolis algorithm grows with the total number of variables.

2.2 Classical liquids

The Metropolis algorithm was first applied in the simulation of the structure of classical liquids and is still used in current research in the study of glass transitions and polymer systems. In this section, we would like to discuss some applications of the Metropolis algorithm in statistical physics. We will use the classical liquid system as the an illustrative example. More detailed accounts of computer simulations of classical liquids can be found in Allen and Tildesley (1987) and Hansen and McDonald (2013). The classical liquid system is continuous in spatial coordinates of atoms and is a good system to study liquid–solid phase transitions, including the situation of glass formation.

Let us assume that the system is in good contact with a thermal bath and has a fixed number of particles and volume size; that is, that the system is described by the canonical ensemble. Then the average of a physical quantity A is given by

$$\langle A \rangle = \frac{1}{Z} \int A(\mathbf{R}) e^{-U(\mathbf{R})/kT} d\mathbf{R}, \tag{2.13}$$

with Z being the partition function of the system, which is given by

$$Z = \int e^{-U(\mathbf{R})/kT} d\mathbf{R}. \tag{2.14}$$

Here $\mathbf{R} = (\mathbf{r}_1, \mathbf{r}_2, \ldots, \mathbf{r}_N)$ is a $3N$-dimensional vector for the coordinates of all the particles in the system and k is Boltzmann's constant. We have suppressed the velocity dependence in the expression with the understanding that the distribution of the velocity is given by the Maxwell distribution, which can be used to calculate the averages of any velocity-related physical quantities. For example, the average kinetic energy component of a particle is given by

$$\left\langle \frac{m v_{ij}^2}{2} \right\rangle = \frac{kT}{2}, \tag{2.15}$$

where $i = 1, 2, \ldots, N$ is the index for the ith particle and $j = 1, 2, 3$ is the index for the directions x, y, and z (Tuckerman 2010). As a matter of fact, the average of any physical quantity associated with velocity can be obtained through the partition theorem. For example, a quantity $B(\mathbf{V})$ with $\mathbf{V} = (\mathbf{v}_1, \mathbf{v}_2, \ldots, \mathbf{v}_N)$ can always be expanded in the Taylor series of the velocities, and then each term can be evaluated with

$$\left\langle v_{ij}^n \right\rangle = \begin{cases} nkT/m & \text{if } n \text{ is even,} \\ 0 & \text{otherwise.} \end{cases} \tag{2.16}$$

Now we can concentrate on the evaluation of a physical quantity that depends on the coordinates only, such as the total potential energy, the pair distribution of the particles, or the pressure in the system. The average of the physical quantity is then given by

$$\langle A \rangle = \frac{1}{Z} \int A(\mathbf{R}) e^{-U(\mathbf{R})/kT} d\mathbf{R} = \frac{1}{M} \sum_{i=1}^{M} A(\mathbf{R}_i), \tag{2.17}$$

with $\mathbf{R}_i$ being a set of points in the configuration space sampled according to the distribution function

$$W(\mathbf{R}) = \frac{e^{-U(\mathbf{R})/kT}}{Z}. \tag{2.18}$$

Assuming that there is a zero external potential; that is $U_{\text{ext}} = 0$, and we have a pairwise interaction between any two particles, the total potential energy for a specific configuration is then given by

$$U(\mathbf{R}) = \sum_{j>i}^{N} U_{\text{int}}(r_{ij}). \tag{2.19}$$

In the preceding section, we discussed how to update all the components in the old configuration in order to reach a new configuration. However, sometimes it is more efficient to update the coordinates of just one particle at a time in the old configuration to reach the new configuration, especially when the system is very close to equilibrium or near a phase transition. Here we would like to show how to update the coordinates of one particle to obtain the new configuration. The coordinates for the ith particle are updated with

$$x_i^{(n+1)} = x_i^{(n)} + h_x(2\alpha_i - 1), \tag{2.20}$$

$$y_i^{(n+1)} = y_i^{(n)} + h_y(2\beta_i - 1), \tag{2.21}$$

$$z_i^{(n+1)} = z_i^{(n)} + h_z(2\gamma_i - 1), \tag{2.22}$$

with h_k being the step size along the kth direction and α_i, β_i, and γ_i being uniform random numbers in the region [0, 1]. The acceptance of the move is determined by importance sampling, that is, by comparing the ratio

$$p = \frac{W(\mathbf{R}_{n+1})}{W(\mathbf{R}_n)} \tag{2.23}$$

with a uniform random number w_i. The attempted move is accepted if $p \geqslant w_i$ and rejected if $p < w_i$. Note that in the new configuration only the coordinates of the ith particle are moved, so we do not need to evaluate the whole $U(\mathbf{R}_{n+1})$ again in order to obtain the ratio p for the pairwise interactions. We can express the ratio in terms of the potential energy difference between the old configuration and the new configuration as

$$p = e^{-\Delta U/kT}, \tag{2.24}$$

with

$$\Delta U = \sum_{j \neq i}^{N} \left[U_{\text{int}}\left(\left| \mathbf{r}_i^{(n+1)} - \mathbf{r}_j^{(n)} \right| \right) - U_{\text{int}}\left(\left| \mathbf{r}_i^{(n)} - \mathbf{r}_j^{(n)} \right| \right) \right], \tag{2.25}$$

which can usually be truncated for particles within a distance $r_{ij} \leqslant r_c$. Here r_c is a typical distance at which the effect of the interaction $U_{int}(r_c)$ is negligible. For example, the interaction in a simple liquid is typically the Lennard-Jones potential, which decreases drastically with the separation of the two particles. The typical distance for the truncation in the Lennard-Jones potential is about $r_c = 3\sigma$, where σ is the separation of the zero potential. As we discussed earlier, we should not take *all* the discrete data points in the sampling for the average of a physical quantity, because the autocorrelation of the data is very high if the points are not far apart. Typically, we need to skip about 10–15 data points before taking another value for the average. The evaluations for physical quantities such as total potential energy, structural factor, and pressure are performed almost exactly the same as in the molecular dynamics if we treat the Monte Carlo steps as the time steps in the molecular dynamics.

Another issue for the simulation of infinite systems is the extension of the finite box with a periodic boundary condition. Long-range interactions, such as the Coulomb interaction, cannot be truncated; a summation of the interaction between particles in all the periodic boxes is needed. The Ewald sum is used to include the interactions between a charged particle and the images in other boxes under the periodic boundary condition. Discussion of the Ewald summation can be found in standard textbooks of solid-state physics, for example, Madelung (1978).

The Monte Carlo step size h is determined from the desired rejection rate. For example, if we want 70% of the moves to be rejected, we can adjust h to satisfy such a rejection rate. A larger h produces a higher rejection rate. In practice, it is clear that a higher rejection rate will produce data points with less fluctuation. However, we also have to consider the computing time needed. The decision is made according to the optimization of the computing time and the accuracy of the data sought. Typically, 50–75% are used as the rejection rate in most Monte Carlo simulations. Here h along each different direction, that is, h_x, h_y, or h_z, is determined according to the length scale along that direction. An isotropic system would have $h_x = h_y = h_z$. A system confined along the z direction, for example, would need a smaller h_z. There have been many Monte Carlo studies of classical liquids, including the formation of the glass state. A good selection of reviews is given in Binder (1995).

2.3 Block algorithms

Now let us turn to discrete model systems. The scheme is more or less the same. The difference now comes mainly from the way the new configuration is attempted. Because the variables are discrete, instead of moving the particles in the system, we need to update the configuration by changing the local state of each lattice site. Here we will use the classical three-dimensional Ising model as an illustrative example. The Hamiltonian is

$$H = -J \sum_{\langle ij \rangle}^{N} s_i s_j - B \sum_{i=1}^{N} s_i, \tag{2.26}$$

where J is the exchange interaction strength, B is the external field, $\langle ij \rangle$ means the summation over all nearest neighbors, and N is the total number of sites in the system. The spin s_i for $i = 1, 2, \ldots, N$ can take values of either 1 or −1, so the summation for the interactions is for energies carried by all the bonds between nearest neighboring sites.

The Ising model was historically used to study magnetic phase transitions. The magnetization is defined as

$$m = \frac{1}{N} \sum_{i=1}^{N} \langle s_i \rangle, \tag{2.27}$$

which is a function of the temperature and external magnetic field. For the $B = 0$ case, there is a critical temperature T_c that separates different phases of the system. For example, the system is ferromagnetic if $T < T_c$, paramagnetic if $T > T_c$, and unstable if $T = T_c$. The complete plot of T, m, and B forms the so-called phase diagram. Another interesting application of the Ising model is that it is also a generic model for binary lattices; that is, two types of particles can occupy the lattice sites with two different on-site energies whose difference is $2B$. So the results obtained from the study of the Ising model apply also to other systems in its class, such as binary lattices. Readers who are interested in these aspects can find good discussions in Parisi (1988). Other quantities of interest include, but are not limited to, the total energy

$$E = \langle H \rangle, \tag{2.28}$$

and the specific heat

$$C = \frac{\langle H^2 \rangle - \langle H \rangle^2}{NkT^2}. \tag{2.29}$$

In order to simulate the Ising model, for example, in the calculation of m, we can apply more or less the same idea for the continuous system. The statistical average of the spin at each site is given by

$$m = \frac{1}{Z} \sum_{\sigma} s_\sigma e^{-H_\sigma/kT}, \tag{2.30}$$

where $s_\sigma = S_\sigma/N$, with $S_\sigma = \sum s_i$ being the total spin of a specific configuration labeled by σ and H_σ being the corresponding Hamiltonian (energy). The summation in equation (2.30) is over all the possible configurations. Here Z is the partition function given by

$$Z = \sum_{\sigma} e^{-H_\sigma/kT}. \tag{2.31}$$

The average of a physical quantity, such as the magnetization, can be obtained from

$$m \simeq \frac{1}{M} \sum_{\sigma=1}^{M} s_\sigma, \tag{2.32}$$

with $\sigma = 1, 2, ..., M$ indicating the configurations sampled according to the distribution function

$$W(S_\sigma) = \frac{e^{-H_\sigma/kT}}{Z}. \tag{2.33}$$

Let us go into some more details for the actual simulations. First we randomly assign 1 and −1 to the spins on all the lattice sites. Then we select one site, which can be picked up either randomly or sequentially. Assume that the ith site is selected to be updated. The update is attempted with the spin at the site reversed; that is,

$$s_i^{(n+1)} = -s_i^{(n)}, \tag{2.34}$$

which is accepted into the new configuration if

$$p = e^{-\Delta H/kT} \geqslant w_i, \tag{2.35}$$

where w_i is a uniform random number in the region [0, 1] and ΔH is the energy difference caused by the reversal of the spin, given by

$$\Delta H = -2J s_i^{(n+1)} \sum_{j=1}^{z} s_j^{(n)}. \tag{2.36}$$

Here j runs through all the nearest neighboring sites of i. Note that the quantity $\sum s_j^{(n)}$ associated with a specific site may be stored and updated during the simulation. Every time, if a spin is reversed and its new value is accepted, we can update the summations accordingly. If the reversal is rejected in the new configuration, no update is needed. The detailed balance condition

$$W(S_\sigma)T(S_\sigma \rightarrow S_\mu) = W(S_\mu)T(S_\mu \rightarrow S_\sigma) \tag{2.37}$$

does not determine the transition rate $T(S_\sigma \rightarrow S_\mu)$ uniquely; therefore we can also use the other properly normalized probability for the Metropolis steps, for example

$$q = \frac{1}{1 + e^{\Delta H/kT}} \tag{2.38}$$

instead of p in the simulation without changing the equilibrium results. The use of q speeds up the convergence at higher temperature. There have been considerable discussions on Monte Carlo simulations for the Ising model or related lattice models, such as percolation models. Interested readers can find these discussions in Binder and Heermann (1988).

There is a practical problem with the Metropolis algorithm in statistical physics when the system under study is approaching a critical point. Imagine that the system is a three-dimensional Ising system. At very high temperature, almost all the spins are uncorrelated, and therefore flipping of a spin has a very high chance of being accepted into the new configuration. So the different configurations can be accessed rather quickly and provide an average close to ergodic behavior. However, when the system is moved toward the critical temperature from above, domains with lower energy start to form. If the interaction is ferromagnetic, we will start to see large clusters with all the spins pointing in the same direction. Now, if only one spin is flipped, the configuration becomes much less favorable because of the increase in energy. The favorable configurations are the ones with all the spins in the domain reversed, a point that it will take a very long time to reach. It means that we need to have all the spins in the domain flipped. This requires a very long sequence of accepted moves. It is known as the *critical slowing down*.

Another way to view it is from the autocorrelation function of the total energy of the system. Because now most steps are not accepted in generating new configurations, the total energy evaluated at each Monte Carlo step is highly correlated. In fact, the relaxation time needed to have the autocorrelation function decrease to near zero, t_c, is proportional to the power of the correlation length ξ, which diverges at the critical point in a bulk system. We have

$$t_c = c\xi^z, \tag{2.39}$$

where c is a constant and the exponent z is called the *dynamical critical exponent*, which is about 2, estimated from standard Monte Carlo simulations. Note that the correlation length is bounded by the size of the simulation box L. Then the relaxation time is

$$t_c = cL^z, \tag{2.40}$$

when the system is very close to the critical point.

The first solution to this problem was devised by Swendsen and Wang (1987). Their block update scheme is based on the nearest-neighbor pair picture of the Ising model, in which the partition function is a result of the contributions of all the nearest pairs of sites in the system. Let us examine closely the effect of a specific pair between sites i and j on the total partition function. We express the partition function of the system in terms of Z_p, the partition function from the rest of pairs with the spins at sites i and j parallel, and Z_f, the corresponding partition function without any restriction on the spins at sites i and j. Then the total partition function of the system is

$$Z = qZ_p + (1 - q)Z_f, \tag{2.41}$$

with

$$q = e^{-4\beta J}, \tag{2.42}$$

which can be interpreted as the probability of having a pair of correlated nearest neighbors decoupled. This is the basis on which Swendsen and Wang devised the block algorithm to remove the critical slowing down.

Here is a summary of their algorithm. We first cluster the sites next to each other that have the same spin orientation. A bond is introduced for any pair of nearest neighbors in a cluster. Each bond is then removed with the probability q. After all the bonds are attempted with some removed and the rest kept, there are more, but smaller, clusters still connected through the remaining bonds. The spins in each small cluster are then flipped all together with a probability of 50%. The new configuration is then used for the next Monte Carlo step. This procedure in fact updates the configuration by flipping blocks of parallel spins, which is similar to the physical process that we would expect when the system is close to the critical point. The speedup in this block algorithm is extremely significant. Numerical simulations show that the dynamical critical exponent is now significantly reduced, with $z \simeq 0.35$ for the two-dimensional Ising model and $z \simeq 0.53$ for the three-dimensional Ising model, instead of $z \simeq 2$ in the spin-by-spin update scheme.

The algorithm of Swendsen and Wang is important but still does not eliminate the critical slowing down completely, because the time needed to reach an uncorrelated energy still increases with the correlation length. An algorithm devised later by Wolff (1989) provides about the same improvement in the two-dimensional Ising model as the Swendsen–Wang algorithm but greater improvement in the three-dimensional Ising model. More interesting, the Wolff algorithm eliminates the critical slowing down completely, that is, $z \simeq 0$, in the four-dimensional Ising model, while the Swendsen–Wang algorithm has $z \simeq 1$.

The idea of Wolff is very similar to the idea of Swendsen and Wang. Instead of removing bonds, Wolff proposed constructing a cluster with nearest sites having the same spin orientation. We first select a site randomly from the system and add its nearest neighbors with the same spin orientation to the cluster with the probability $p = 1 - q$. This is continued, with sites added to the cluster, until all the sites in the system are attempted. All the spins in the constructed cluster are then flipped together to reach a new spin configuration of the system. We can easily show that the Wolff algorithm ensures detailed balance.

The Swendsen–Wang and Wolff algorithms can also be generalized to other spin models, such as the xy model and the Heisenberg model (Wolff 1989). The idea comes from the construction of the probability q in both of the above-mentioned algorithms. At every step, we first generate a random unit vector $\mathbf{e}$ with each component a uniform random number x_i,

$$\mathbf{e} = \frac{1}{r}(x_1, x_2, \ldots, x_n), \qquad (2.43)$$

with

$$r^2 = \sum_{i=1}^{n} x_i^2, \qquad (2.44)$$

where n is the dimension of the vector space of the spin. We can then define the probability of breaking a bond as

$$q = \min \left\{ 1, \ e^{-4\beta J (\mathbf{e} \cdot \mathbf{s}_i)(\mathbf{e} \cdot \mathbf{s}_j)} \right\}, \tag{2.45}$$

which reduces to equation (2.42) for the Ising model. The Swendsen–Wang algorithm and the Wolff algorithm have also shown significant improvement in the xy model, the Heisenberg model, and the Potts model. For more details, see Swendsen *et al* (1992).

Exercises

2.1 Show that the Monte Carlo quadrature yields a standard deviation

$$\Delta^2 = \langle A^2 \rangle - \langle A^2 \rangle \propto \frac{1}{M},$$

where A is a physical observable and M is the total number of Monte Carlo points taken. Demonstrate it numerically by sampling the average of a set of data $x_i \in [0, 1]$, drawn uniformly from a random-number generator.

2.2 Show that the Metropolis algorithm applied to statistical mechanics does satisfy the detailed balance and sample the points according to the distribution function. Demonstrate it by sampling the speed of a particle in an ideal gas.

2.3 Calculate the integral

$$S = \int_{-\infty}^{\infty} e^{-r^2/2} (xyz)^2 \ d\mathbf{r}$$

with the Metropolis algorithm and compare the Monte Carlo result with the exact result. Does the Monte Carlo errorbar decrease with the total number of points as expected?

2.4 Calculate the autocorrelation function of the data sampled in the evaluation of the integral in exercise 2.3. From the plot of the autocorrelation function, determine how many points need to be skipped between any two data points in order to have nearly uncorrelated data points.

2.5 Develop a Monte Carlo program for the ferromagnetic Ising model on a square lattice. For simplicity, the system can be chosen as an $N \times N$ square with the periodic boundary condition imposed. Update the spin configuration site by site. Study the temperature dependence of the magnetization. Is there any critical slowing down in the simulation when the system is approaching the critical point?

2.6 Implement the Swendsen–Wang algorithm in the Monte Carlo study of the ferromagnetic Ising model on a triangular lattice. Does the Swendsen–Wang scheme cure the critical slowing down completely?

2.7 Implement the Wolff algorithm in the Monte Carlo study of the ferromagnetic Heisenberg model on a cubic lattice. Does the Wolff scheme cure the critical slowing down completely?

2.8 Study the antiferromagnetic Ising model on a square lattice with both the Swendsen–Wang and Wolff algorithms. Find the temperature dependence of the staggered magnetization. Which of the algorithms handles the critical slowing down better? What happens if the system is a triangular lattice?

2.9 Carry out the Monte Carlo study of the anisotropic Heisenberg model

$$H = -J \sum_{\langle ij \rangle}^{L} \left(\lambda s_i^z s_j^z + s_i^x s_j^x + s_i^y s_j^y \right)$$

on a square lattice, where $J > 0$ and the spins are classical, each with a magnitude S. Find the λ dependence of the critical temperature for a chosen value of S. What happens if $J < 0$?

Bibliography

Allen M P and Tildesley D J 1987 *Computer Simulation of Liquids* (London: Clarendon Press)

Binder K 1995 *The Monte Carlo Method in Condensed Matter Physics* (Berlin: Springer)

Binder K and Heermann D W 1988 *Monte Carlo Simulation in Statistical Physics, An Introduction* (Berlin: Springer)

Hansen J-P and McDonald I R 2013 *Theory of Simple Liquids* (San Diego, CA: Academic)

Madelung O 1978 *Introduction to Solid-State Theory* (Berlin: Springer)

Metropolis N, Rosenbluth A W, Rosenbluth M N, Teller A H and Teller E 1953 Equation of state calculations by fast computing machines *J. Chem. Phys.* **21** 1087–92

Pang T 2006 *An Introduction to Computational Physics* 2nd ed (Cambridge: Cambridge University Press) ch 10

Parisi G 1988 *Statistical Field Theory* (Redwood City, CA: Addison-Wesley)

Swendsen R H and Wang J-S 1987 Nonuniversal critical dynamics in Monte Carlo simulations *Phys. Rev. Lett.* **58** 86–8

Swendsen R H, Wang J-S and Ferrenberg A M 1992 New Monte Carlo methods for improved efficiency of computer simulation in statistical mechanics *The Monte Carlo Methods in Condensed Matter Physics* ed K Binder (Berlin: Springer) pp 75–91

Tuckerman M E 2010 *Statistical Mechanics: Theory and Molecular Simulation* (Oxford: Oxford University Press)

Wolff U 1989 Collective Monte Carlo updating for spin systems *Phys. Rev. Lett.* **62** 361–4

Chapter 3

Variational Monte Carlo

The most direct application of the Metropolis algorithm to quantum many-body systems is to combine it with the variational principle. Consider a quantum system of N interacting, identical particles, each of mass m. The Hamiltonian of this many-body system is given by

$$H = K + U, \tag{3.1}$$

where the kinetic energy operator is

$$K = -\frac{\hbar^2}{2m} \sum_{i=1}^{N} \nabla_i^2 \tag{3.2}$$

and the potential energy operator is

$$U = \sum_{i=1}^{N} U_{\text{ext}}(\mathbf{r}_i) + \sum_{i>j=1}^{N} U_{\text{int}}(r_{ij}). \tag{3.3}$$

We have made several assumptions in the above expressions: the particles are identical with the same mass m, the external potential energy on each particle is a function of its position, and the interactions between particles are pairwise. The corresponding time-independent Schrödinger equation can symbolically be written as

$$H\Psi_n(\mathbf{R}) = E_n\Psi_n(\mathbf{R}), \tag{3.4}$$

where $\Psi_n(\mathbf{R})$ and E_n are the nth eigenstate and the corresponding eigenvalue of H, respectively (Schrödinger 1926). The purpose of a variational approach is to find a wavefunction that can be optimized to produce a closest expectation value to the ground-state energy of the system.

doi:10.1088/978-1-6817-4109-3ch3

3.1 Variational principle

Usually an analytical or exact solution of the Schrödinger equation of a system with more than two particles is hard to find. Therefore, approximate methods are important tools for studying quantum systems. It is easiest to study the ground state of a many-body system in comparison with other states of the same system.

We can use a trial state $\Phi(\mathbf{R})$ to approximate the ground-state wavefunction. The parameters or functions in $\Phi(\mathbf{R})$ are then optimized according to the variational principle:

$$E[\alpha_i] = \frac{\langle \Phi | H | \Phi \rangle}{\langle \Phi | \Phi \rangle} \geqslant E_0, \tag{3.5}$$

where E_0 is the ground-state energy of the system and α_i is a set of linearly independent variational parameters or functions. This inequality can be shown by expanding $\Phi(\mathbf{R})$ in terms of the eigenstates of the Hamiltonian with

$$\Phi(\mathbf{R}) = \sum_{n=0}^{\infty} a_n \Psi_n(\mathbf{R}). \tag{3.6}$$

The expansion above is a generalization of the Fourier theorem because the $\Psi_n(\mathbf{R})$ form a complete basis set. The variational principle results if we substitute the expansion for $\Phi(\mathbf{R})$ into the expectation value with the understanding that $E_n \geqslant E_0$ for $n > 0$ and $\langle \Psi_n | \Psi_m \rangle = \delta_{nm}$; namely,

$$E[\alpha_i] = \frac{\displaystyle\sum_{n=0}^{\infty} a_n^2 E_n}{\displaystyle\sum_{n=0}^{\infty} a_n^2} \geqslant E_0. \tag{3.7}$$

See Pang (2006) for more discussion on the Fourier expansion and its convergence.

3.2 The Metropolis step

To simplify our discussion, we assume that the space hosting the system is a continuum and that α_i is a set of variational parameters. Hence, we can write the expectation value in the form of an integral

$$E[\alpha_i] = \frac{\displaystyle\int \Phi^{\dagger}(\mathbf{R}) H \Phi(\mathbf{R}) \, d\mathbf{R}}{\displaystyle\int |\Phi(\mathbf{R}')|^2 \, d\mathbf{R}'} = \int W(\mathbf{R}) E(\mathbf{R}) \, d\mathbf{R}, \tag{3.8}$$

where

$$W(\mathbf{R}) = \frac{|\Phi(\mathbf{R})|^2}{\displaystyle\int |\Phi(\mathbf{R}')|^2 \, d\mathbf{R}'} \tag{3.9}$$

can be interpreted as a distribution function and

$$E(\mathbf{R}) = \frac{H\Phi(\mathbf{R})}{\Phi(\mathbf{R})} \qquad (3.10)$$

can be viewed as a local energy, a physical quantity of the system at configuration $\mathbf{R}$. The expectation value $E[\alpha_i]$ can be evaluated if the expressions for $E(\mathbf{R})$ and $W(\mathbf{R})$ for a specific set of parameters α_i are available. In practice, $\Phi(\mathbf{R})$ can be para-meterized to include the relevant physics. The variational parameters α_i in the trial wavefunction $\Phi(\mathbf{R})$ can then be optimized by minimizing $E[\alpha_i]$.

Each evaluation of the expectation value in the variational quantum Monte Carlo method for a given set of variational parameters is equivalent to the evaluation of an average of a physical quantity of a classical system with the same Metropolis algorithm for a given temperature. During the sampling process, we can either update the entire configuration or just the coordinates associated with a particular particle at each step. If the variational wavefunction $\Phi(\mathbf{R})$ is sufficiently simple, for example, it contains a small number of parameters, we can search for the optimal state by varying each parameter in the simulation. If the form of the wavefunction is more complicated, or there are more than a few parameters, a systematic scheme can be designed to optimize the wavefunction based on the Euler–Lagrange equation

$$\frac{\delta E[\alpha_i]}{\delta \alpha_j} = 0, \qquad (3.11)$$

for $j = 1, 2, 3, ...$ Good examples of performing a systematic search of variational parameters are available (Umrigar et al 1988). It is important to include in the variational wavefunction a way to cope with the divergence of the interaction when two particles approach each other. It is common to have this divergence canceled by the relative kinetic energy of the two particles when their separation goes to zero. Such a constraint on the wavefunction is called the *cusp condition* (Mahan 2000) and can be found by solving the two-body problem analytically for a vanishing separation distance. If the cusp condition is built into the variational wavefunction, fluctuations of the results are significantly reduced, and we can simulate a system with a larger number of particles and greater accuracy. We will demonstrate this point in the guide wavefunction in the example of ^{4}He clusters trapped on the surface of graphite in the next chapter.

Here we would like to demonstrate how to use the Metropolis algorithm to move all the coordinates of the particles in the system.

```
// Method to make one Metroplolis move for all the particles.

    public static void metropolis() {

        double cfsv[] = new double[nv];

    // Take an atempted step for every coordinate
```

```
for (int i=0; i<N; ++i) {
  cfsv[i] = conf[i];
  conf[i] += (ranf()-0.5)*dx;
  cfsv[N+i] = conf[N+i];
  conf[N+i] += (ranf()-0.5)*dy;
  cfsv[2*N+i] = conf[2*N+i];
  conf[2*N+i] += (ranf()-0.5)*dz;
}

double wold = w;
wave();
++ia;

// Reassign previous coordinates if the move is rejected

if (Math.exp(w-wold)<ranf()) {
  for (int i=0; i<nv; ++i) conf[i] = cfsv[i];
  w = wold;
  --ia;
}
}
```

Note that the weight used in the above is calculated in the method that evaluates the wavefunction for the new configuration.

3.3 Kinetic energy and wavefunction

The kinetic energy of the ith particle, K_i, can be written in two parts as

$$K_i = -\frac{\hbar^2}{2m}\frac{\nabla_i^2 \Phi(\mathbf{R})}{\Phi(\mathbf{R})} = 2T_i - |\mathbf{F}_i|^2, \tag{3.12}$$

where

$$T_i = -\frac{\hbar^2}{4m}\nabla_i^2 \ln|\Phi(\mathbf{R})| \tag{3.13}$$

and

$$\mathbf{F}_i = \frac{\hbar}{\sqrt{2m}}\nabla_i \ln|\Phi(\mathbf{R})|. \tag{3.14}$$

In principle, both T_i and $|\mathbf{F}_i|^2$ converge to the kinetic energy K_i, which can be shown by integration by parts (Jackson and Feenberg 1961). In practice, each term fluctuates during sampling. The combination minimizes the fluctuations because the two terms have opposite signs. The separate evaluation of T_i and $|\mathbf{F}_i|^2$ also provides an independent check on the convergence and validity of the average kinetic energy through the combination.

A common choice of the trial wavefunction for quantum liquids has the general form

$$\Phi(\mathbf{R}) = D(\mathbf{R})e^{-U_J(\mathbf{R})}, \tag{3.15}$$

where $D(\mathbf{R})$ is a constant for boson systems and a Slater determinant of single-particle orbitals for fermion systems to have the Pauli principle met. Here $U_J(\mathbf{R})$ is the Jastrow correlation factor, which can be written in terms of one-body terms, two-body terms, and so on, with

$$U_J(\mathbf{R}) = \sum_{i=1}^{N} u_1(\mathbf{r}_i) + \sum_{i>j}^{N} u_2(\mathbf{r}_i, \mathbf{r}_j) + \sum_{i>j>k}^{N} u_3(\mathbf{r}_i, \mathbf{r}_j, \mathbf{r}_k) + \cdots, \tag{3.16}$$

which is usually truncated at the two-body terms in most Monte Carlo simulations. Both $u_1(\mathbf{r})$ and $u_2(\mathbf{r}, \mathbf{r}')$ are parameterized according to the physical understanding of the source of these terms We can show that when the external potential becomes a dominant term, $u_1(\mathbf{r})$ is uniquely determined by the form of the external potential, and when the interaction between the ith and jth particles dominates, the form of $u_2(\mathbf{r}_i, \mathbf{r}_j)$ is uniquely determined by $U_{\text{int}}(\mathbf{r}_i, \mathbf{r}_j)$. These are the important aspects used in determining the form of u_1 and u_2 in practice.

First let us consider the case of bulk helium liquids. Because the external potential is zero, we can choose $u_1(\mathbf{r}) = 0$. The two-body term is translationally invariant, $u_2(\mathbf{r}_i, \mathbf{r}_j) = u_2(r_{ij})$. The expression of $u_2(r)$ at the limit of $r \to 0$ can be obtained by solving a two-body problem. In the center-of-mass coordinate system, we have

$$\left[-\frac{\hbar^2}{2\mu}\nabla^2 + U_{\text{int}}(r) \right]e^{-u_2(r)} = Ee^{-u_2(r)}, \tag{3.17}$$

where μ is the reduced mass $m/2$, $U_{\text{int}}(r)$ is the interaction potential that is given by $4\varepsilon(\sigma/r)^{12}$ at the limit of $r \to 0$, and the Laplacian can be decoupled in the angular momentum eigenstates as

$$\nabla^2 = \frac{d^2}{dr^2} + \frac{2}{r}\frac{d}{dr} - \frac{l(l+1)}{r^2}. \tag{3.18}$$

We can show that in order to have the divergence in the potential energy canceled by the kinetic energy at the limit of $r \to 0$, we must have

$$u_2(r) = \left(\frac{a}{r} \right)^5. \tag{3.19}$$

The condition needed to remove the divergence of the potential energy with the kinetic energy term constructed from the wavefunction is the cusp condition mentioned earlier. This condition is extremely important in all quantum Monte

Carlo simulations, because it is the major means of stabilizing the algorithms. The behavior of $u_2(r)$ at longer range is usually dominated by the density fluctuation or zero-point motion of phonons, which is proportional to $1/r^2$. We can also show that the three-body term and the effects due to the backflow are also important and can be incorporated into the variational wavefunction. The Slater determinant for liquid ^{3}He can be constructed from plane waves. The key, as we have stressed earlier, is to find a variational wavefunction that contains the essential physics of the system, which is always nontrivial. An interesting work by Bouchaud and Lhuillier (1987) replaced the Slater determinant with a Bardeen–Cooper–Schrieffer (BCS) ansatz for liquid ^{3}He. This changes the structure of the ground state of the system from a Fermi liquid, which is characterized by a Fermi surface with a finite jump in the particle distribution, to a superconducting glass, which does not have any Fermi surface.

Electronic systems are another class of systems studied with the variational quantum Monte Carlo method. Significant results are obtained for atomic systems, molecules, and even solids. Typical approaches treat electrons and nuclei separately with the so-called Born–Oppenheimer approximation (Born and Oppenheimer 1927); that is, the electronic state is adjusted quickly for a given nuclear config- uration so the potential due to the nuclei can be treated as an external potential of the electronic system. Assume that we can obtain the potential of the nuclei or ions with some other methods. Then $u_1(\mathbf{r}_i)$ can be parameterized to ensure the cusp condition between an electron and a nucleus or ion. For the ion with an effective charge Ze, we have

$$u_1(r) = Zr/a_0 \qquad (3.20)$$

as $r \to 0$. Here a_0 is the Bohr radius. $u_2(r_{ij})$ is obtained from the electron–electron interaction, and the cusp condition requires

$$u_2(r_{ij}) = -\frac{\sigma_{ij} r_{ij}}{2a_0} \qquad (3.21)$$

as $r_{ij} \to 0$. Here $\sigma_{ij} = 1$ if the ith and jth electrons have different spin orientations; otherwise $\sigma_{ij} = 1/2$. The Slater determinant can be constructed from the local orbitals, for example, the linear combinations of the Gaussian orbitals from all nuclear sites.

There have been a lot of impressive variational quantum Monte Carlo simu- lations for electronic systems, for example, the work of Fahy *et al* (1990) on carbon and silicon solids, the work of Umrigar *et al* (1988) on the improvement of variational wavefunctions, and the work of Umrigar (1993) on accelerated varia- tional Monte Carlo simulations.

The numerical procedure for performing variational quantum Monte Carlo simulations is exactly the same as that for the Monte Carlo simulations of statistical systems. The only difference is that for the statistical system the calculations are done for a given temperature, but for the quantum system they are done for a given set of variational parameters in the variational wavefunction. We can update either the whole configuration or just the coordinates associated with a particular particle at each Metropolis step.

3.4 Quantum dots

As an illustrative example, let us consider a circular quantum dot with N electrons trapped in a harmonic potential $U_{ext}(r) = m\omega_0^2 r^2/2$, where m is the electron mass, ω_0 is the corresponding angular frequency of the potential, and $r = \sqrt{x^2 + y^2}$ is the distance from the center of the potential. The total Hamiltonian of the system is then given by

$$H = K + U, \tag{3.22}$$

where the kinetic energy is

$$K = -\frac{\hbar^2}{2m} \sum_{i=1}^{N} \nabla_i^2 \tag{3.23}$$

and the potential energy is

$$U = \sum_{i=1}^{N} U_{ext}(r_i) + \frac{e^2}{4\pi\epsilon_0} \sum_{i>j=1}^{N} \frac{1}{r_{ij}}, \tag{3.24}$$

where r_{ij} is the distance between the ith and jth electrons.

A simple trial wavefunction for this system can be constructed from two Slate determinants, one for spin-up and the other for spin-down electrons, of the single-particle states of the two-dimensional harmonic oscillator:

$$\phi_{n_x, n_y}(x, y) = \frac{\gamma}{\sqrt{\pi 2^{n_x + n_y} n_x! n_y!}} e^{-\gamma^2(x^2 + y^2)/2} H_{n_x}(\gamma x) H_{n_y}(\gamma y), \tag{3.25}$$

where $H_n(z)$ is the nth-order Hermite polynomial, and the Jastrow correlation factor between any two electrons mentioned earlier.

Recently, quantum Monte Carlo simulations of electronic systems, including two-dimensional harmonic quantum dots, have also incorporated the spin–orbit interactions (Melton *et al* 2016), which lead to the formation of spin textures in the systems.

Exercises

3.1 Derive the relation of equation (3.12) by calculating the expectation values of all the terms and show that the expectation values of both T_i and $|\mathbf{F}_i|^2$ converge to that of K_i after performing an integration by parts.

3.2 Find the ground-state energy of a three-dimensional harmonic oscillator with the Hamiltonian $H = -(\hbar^2/2m)\nabla^2 + kr^2/2$, using the variational Monte Carlo method with the trial/guide wavefunction $\phi(r) = e^{-r^2/2}$ and natural units of $m = k = \hbar = 1$. Compare your simulation results with the exact result, $E_0 = 3/2$.

3.3 Find the ground-state energy of the hydrogen molecule using the variational Monte Carlo method. Assume that the protons are stationary with a fixed separation of 0.741 Å. A possible choice of the trial/guide wavefunction of the two electrons is $\Phi(\mathbf{r}_1, \mathbf{r}_2) = \phi(\mathbf{r}_1)\phi(\mathbf{r}_2)e^{r_{12}/(2a_0)}$, with $\phi(\mathbf{r})$ a linear combination of the 1s atomic orbitals from the two protons and a_0 is the Bohr radius. Show that the cusp condition is satisfied between the two electrons and between an electron and a proton. Improve the trial wavefunction with variational parameters or variations of the functional form while maintaining the cusp condition. Compare the calculated result with the observed dissociation energy of the molecule, 4.52 eV.

3.4 Estimate the ground-state energy of the lithium atom using the variational Monte Carlo method. One choice of the trial/guide wavefunction is to combine the 1s and 2s atomic orbitals and a Jastrow factor. Use a Slater determinant for the two electrons with the same spin and reject any time step that causes the system to cross a node of the wavefunction. Compare the calculated result with the observed ground state energy of the atom, -203.4860 eV.

3.5 Calculate the ground-state energy per atom and pair-distribution function of bulk ^{4}He using the variational Monte Carlo method. Enforce periodic boundary conditions on the system by moving a particle back to the simulation box on the opposite side when a move takes it to the outside of the box. Use the Aziz potential given in Aziz *et al* (1995) for the interaction between the atoms, $\rho = 0.024494$ Å^{-3} for the density, and $N = 32$ atoms in the simulation box with the starting positions on a face-centered cubic lattice.

3.6 Simulate a system of hard-sphere bosons in a three-dimensional harmonic potential well $U_{ext}(r) = kr^2/2$. The model can be viewed as a simple model of cold atoms in a trap. Use $\Phi(\mathbf{R}) = \prod_{i=1}^{N} e^{-r_i^2}$ as the trial/guide and reject any move that would cause the separation of any two particles r_{ij} to be less than d, where d is the diameter of the hard spheres. Calculate E_0/N for various values of N with natural units of $m = k = \hbar = 1$. Check your result in the limit $d \to 0$ against $E_0/N = 3/2$. Start the simulation with N on the order of 10 and increase it to 1000. What is the N dependence of the computation time if the same accuracy of calculation is sought?

3.7 Study the electronic structure of the helium atom with the variational quantum Monte Carlo method. Assume that the nucleus is fixed at the origin of the coordinates. The key is to find a good parameterized variational wavefunction with proper cusp conditions built in.

3.8 Find the variational ground-state energy per particle, the density profile, and the pair correlation function of a ^{4}He cluster. How sensitive are the values to the size of the cluster? Assume that the interaction between any two atoms is given by the Lennard-Jones potential and express the results in terms of the potential parameters. What happens if the system is a ^{3}He cluster?

Bibliography

Aziz R A, Janzen A R and Moldover M R 1995 Ab initio calculations for helium: A standard for transport property measurements *Phys. Rev. Lett.* **74** 1586–9

Born M and Oppenheimer J R 1927 Zur quantentheorie der molekeln *Annalen der Physik* **389** 457–84

Bouchaud J P and Lhuillier C 1987 A new variational description of liquid ^{3}He: the superfluid glass *Europhys. Lett.* **3** 1273–80

Fahy S, Wang X W and Louie S G 1990 Variational quantum Monte Carlo nonlocal pseudopotential approach to solids: Formulation and application to diamond, graphite, and silicon *Phys. Rev.* B **42** 3503–22

Jackson H W and Feenberg E 1961 Perturbation method for low states of a many-particle boson system *Ann. Phys.* **15** 266–95

Mahan G 2000 *Many Particle Physics* 3rd ed (Berlin: Springer) ch 10

Melton C A, Bennett M C and Mitas L 2016 Quantum Monte Carlo with variable spins *The J. Chem. Phys.* **144** 244113

Pang T 2006 *An Introduction to Computational Physics* 2nd ed (Cambridge: Cambridge University Press) ch 6

Schrödinger E 1926 An undulatory theory of the mechanics of atoms and molecules *Phys. Rev.* **28** 104970

Umrigar C J 1993 Accelerated Metropolis method *Phys. Rev. Lett.* **71** 408–11

Umrigar C J, Wilson K G and Wilkins J W 1988 Optimized trial wave functions for quantum Monte Carlo calculations *Phys. Rev. Lett.* **60** 1719–22

An Introduction to Quantum Monte Carlo Methods

Tao Pang

Chapter 4

Diffusion Monte Carlo

The limitation of the variational quantum Monte Carlo method is obvious. Unless we know the exact form of the ground-state wavefunction, the simulation will never reach the true ground state. In many cases, we may not even have enough information to construct a decent variational wavefunction. Is there still a way to simulate a many-body system if we have limited knowledge? The answer is the diffusion quantum Monte Carlo method. It is an exact method for the ground state of many-boson systems and yields a high potential in probing many-fermion systems as well. The strategy of the diffusion Monte Carlo is to map the time-dependent Schrödinger equation into an imaginary-time diffusion equation to filter out the excited states of the system over time through a diffusion process (Anderson 2007).

4.1 The algorithm

We can map the time-dependent Schrödinger equation

$$i\hbar \frac{\partial \Psi(\mathbf{R}, t)}{\partial t} = H \Psi(\mathbf{R}, t) \tag{4.1}$$

into a diffusion equation with imaginary-time as

$$\frac{\partial \Psi(\mathbf{R}, \tau)}{\partial \tau} = -(H - E_R)\Psi(\mathbf{R}, \tau), \tag{4.2}$$

where E_R is a reference energy that can be interpreted as the zero point of the system's energy at the given moment and the imaginary time $\tau = it/\hbar$, with t being the real time from the original Schrödinger equation. Here E_R is adjusted at each time step in the actual simulation so that the simulation will converge faster. We have absorbed Planck's constant $\hbar$ into the time for convenience. The solution of the diffusion equation (4.2) can symbolically be written as

$$\Psi(\mathbf{R}, \tau) = e^{-(H-E_{\mathrm{R}})\tau}\Psi(\mathbf{R}, 0) \tag{4.3}$$

if we start from an initial state $\Psi(\mathbf{R}, 0)$.

A good choice for the initial configuration is $\Psi(\mathbf{R}, 0) = \Phi(\mathbf{R})$, the trial wavefunction that we have already found from the variational Monte Carlo simulation. We can multiply the imaginary-time Schrödinger equation by $\Phi(\mathbf{R})$ and rewrite it as a diffusion equation for $f(\mathbf{R}, \tau) = \Phi(\mathbf{R})\Psi(\mathbf{R}, \tau)$:

$$\frac{\partial f(\mathbf{R}, \tau)}{\partial \tau} = \frac{\hbar^2}{2m}\nabla^2 f(\mathbf{R}, \tau) - \nabla \cdot [\mathbf{V}(\mathbf{R})f(\mathbf{R}, \tau)] - [E(\mathbf{R}) - E_{\mathrm{R}}]f(\mathbf{R}, \tau), \tag{4.4}$$

where the local energy $E(\mathbf{R})$ was defined earlier and $f(\mathbf{R}, \tau)$ can be interpreted as a distribution function of the system with configuration $\mathbf{R}$ at time τ if $f(\mathbf{R}, \tau)$ is always positive. The vector

$$\mathbf{V}(\mathbf{R}) = \frac{\hbar^2}{m}\nabla \ln|\Phi(\mathbf{R})| = \sqrt{\frac{2\hbar^2}{m}}\,\mathbf{F} \tag{4.5}$$

can be interpreted as a drift velocity of the distribution in configuration space.

It is clear from the form of equation (4.4) that the time evolution of $f(\mathbf{R}, \tau)$ is influenced by three contributions, a pure diffusion term involving ∇^2, a drift term involving $\mathbf{V}$, and a source/sink term involving $E - E_{\mathrm{R}}$.

The short-time evolution of $f(\mathbf{R}, \tau)$ is given by

$$f(\mathbf{R}', \tau + \Delta\tau) = \int G(\mathbf{R}', \mathbf{R}; \Delta\tau)f(\mathbf{R}, \tau)\mathrm{d}\mathbf{R}, \tag{4.6}$$

where the propagator

$$G(\mathbf{R}', \mathbf{R}; \Delta\tau) = \left(\frac{1}{2\pi\chi^2}\right)^{3N/2} e^{-\left[\mathbf{R}'-\mathbf{R}-\Delta\tau\mathbf{V}(\mathbf{R})\right]^2\big/2\chi^2 - \Delta\tau[E(\mathbf{R})-E_{\mathrm{R}}]} \tag{4.7}$$

carries the system from configuration $\mathbf{R}$ at time τ to the new configuration $\mathbf{R}'$ at time $\tau + \Delta\tau$.

The three contributions can be clearly identified in $G(\mathbf{R}', \mathbf{R}; \Delta\tau)$. The first is a random walk, following a Gaussian distribution with variance $\chi^2 = \hbar^2\Delta\tau/m$. The second is a drift process that moves the mean value of the configuration by $\Delta\tau\mathbf{V}(\mathbf{R})$. To include these two contributions, the configuration is updated according to

$$\mathbf{R}' = \mathbf{R} + \Delta\tau\mathbf{V}(\mathbf{R}) + \boldsymbol{\xi}, \tag{4.8}$$

where $\boldsymbol{\xi}$ is a $3N$-dimensional Gaussian random number with a variance of χ^2 in each component.

What is left in the propagator is a contribution which either lowers or raises the population of configurations in the ensemble according to the birth/death rate

$$W_{\mathrm{B}}(\mathbf{R}) \propto e^{-\Delta\tau[E(\mathbf{R})-E_{\mathrm{R}}]}. \tag{4.9}$$

The lower the value of the local energy $E(\mathbf{R})$, the higher the rate. This part of the contributions in the propagator can be sampled by a *branching* process: $M_B \propto W_B(\mathbf{R})$ copies of the configuration $\mathbf{R}$ are created to be part of the new ensemble and a mechanism, which does not influence the relative weight of each configuration, needs to be devised to control the overall number of configurations in the ensemble. Usually this control is done by adjusting the reference energy

$$E_R \rightarrow E_R + \kappa \ln \frac{N_P}{N_E} \qquad (4.10)$$

during the simulation with N_P being the preferred population targeted from a current population of N_E. The small parameter κ is selected to control the speed of adjustment. The adjusted reference energy approaches the ground-state energy of the system as the simulation progresses.

4.2 Evaluation of physical quantities

The sampling process starts from a selected trial wavefunction, $\Phi(\mathbf{R})$, which evolves through the diffusion process by the propagator to the ground state of the system. For a boson system, the function $f(\mathbf{R}, \tau)$ can always be made real and positive and behaves as a true distribution. Simulation errors still arise from the finite size of the system, the finite time step, and the finite number of samples. In principle, the guide wavefunction does not affect the outcome of the simulation if the simulation is performed for a long time and if the guide wavefunction has a finite overlap with the actual ground state of the system. In practice, the overlap between the guide wavefunction and the true ground state impacts the speed of convergence of the simulation, and therefore is an important aspect to consider. For a fermion system, the wavefunction has a nodal structure that is usually unknown and further approximations have to be made because of the fermion-sign problem.

Energy

Although the reference energy approaches the ground-state energy of the system if the simulation converges, we do not use it to estimate the ground-state energy because it is adjusted during the simulation. Instead the ground-state energy is sampled from the distribution function $f(\mathbf{R}, \tau)$ at large time τ. Because

$$\lim_{\tau \to \infty} f(\mathbf{R}, \tau) = \Phi(\mathbf{R})\Psi_0(\mathbf{R}), \qquad (4.11)$$

the ground-state energy can be obtained from the time-dependent variational energy

$$E(\tau) = \frac{\langle \Phi|H|\Psi \rangle}{\langle \Phi|\Psi \rangle} = \frac{\int E(\mathbf{R})f(\mathbf{R}, \tau)\,d\mathbf{R}}{\int f(\mathbf{R}, \tau)\,d\mathbf{R}}, \qquad (4.12)$$

which can be interpreted as an average sampled over the distribution function $f(\mathbf{R}, \tau)$. For example, if we have an ensemble of N configurations $\mathbf{R}_1, \mathbf{R}_2, \ldots, \mathbf{R}_{N_E}$, distributed according to $f(\mathbf{R}, \tau)$, the average energy is given by

$$E(\tau) = \frac{\sum_{i=1}^{N_E} E(\mathbf{R}_i)\, W_B(\mathbf{R}_i)}{\sum_{i=1}^{N_E} W_B(\mathbf{R}_i)}, \tag{4.13}$$

which becomes the exact ground-state energy E_0 at infinite time as $f(\mathbf{R}, \tau)$ approaches its limiting value given by equation (4.11). Our discussion forms the basis of the diffusion Monte Carlo for a boson system and can be refined by some other steps. The update in the drift part is accurate only up to the linear term in $\Delta\tau$ and can be improved with a better approximation of the short-time propagator (Forbert and Chin 2001). For example, the second-order approximation of the propagator replaces $\Delta\tau V(\mathbf{R})$ by $\Delta\tau V(\mathbf{R}_t)$ (Astrakharchik 2004), where

$$\mathbf{R}_t = \mathbf{R} + \frac{\Delta\tau}{4}\left\{V(\mathbf{R}) + V\left[\mathbf{R} + \frac{\Delta\tau}{2}V(\mathbf{R})\right]\right\}. \tag{4.14}$$

To reduce the branching fluctuations, we can replace the local energy $E(\mathbf{R})$ in $W_B(\mathbf{R})$ by the average energy of the old and new configurations, $[E(\mathbf{R}) + E(\mathbf{R}')]/2$, resulting in a smoother propagator. Also, a Metropolis move can be inserted between the configuration update and branching with the acceptance probability

$$p = \min\left\{1, \frac{|\Phi(\mathbf{R}')|^2 G(\mathbf{R}, \mathbf{R}'; \Delta\tau)}{|\Phi(\mathbf{R})|^2 G(\mathbf{R}', \mathbf{R}; \Delta\tau)}\right\} > \eta \tag{4.15}$$

for the new configuration $\mathbf{R}'$. This additional step ensures detailed balance between $\mathbf{R}$ and $\mathbf{R}'$ in configuration space.

For a fermion system, a boson-equivalent approach, called the *fixed-node approximation* (Reynolds *et al* 1982) can be introduced to reject any attempt that would cross a node in the trial wavefunction. Several methods have been introduced to relax the nodes of the guide wavefunction, with only limited success, due in part to the intrinsic complexity of the fermion-sign problem (Troyer and Wiese 2005).

Structural analysis

We can analyze the structural properties of a many-body system from the behavior of its distribution functions. Suppose that a system has N particles and is in its ground state described by the wavefunction $\Psi_0(\mathbf{R})$. The n-body distribution function is given by

$$\rho_n(\mathbf{r}_1, \mathbf{r}_2, \ldots, \mathbf{r}_n) = \frac{1}{Z}\frac{N}{(N-n)}\int |\Psi_0(\mathbf{R})|^2\, d\mathbf{r}_{n+1}\, d\mathbf{r}_{n+2}\ldots d\mathbf{r}_N, \tag{4.16}$$

where

$$Z = \int |\Psi_0(\mathbf{R})|^2\, d\mathbf{R} \tag{4.17}$$

is the normalization factor. One of the most interesting distribution functions is the pair-distribution function $g(\mathbf{r}, \mathbf{r}')$, which is defined by

$$\rho_1(\mathbf{r})g(\mathbf{r}, \mathbf{r}')\rho_1(\mathbf{r}') = \rho_2(\mathbf{r}, \mathbf{r}') \tag{4.18}$$

and contains information about long-range order in the system.

For a uniform system, $\rho_1(\mathbf{r})$ is the density of the system, $\rho = N/V$, where V is the volume of the system, and $g(\mathbf{r}, \mathbf{r}') = g(|\mathbf{r}' - \mathbf{r}|)$. The radial distribution function $g(r)$ can be interpreted as the probability of finding another particle a distance r away from a given particle. If we know $g(r)$, we can find the corresponding structure function, $S(k)$, which is measured in a typical scattering experiment. Formally $S(k) - 1$ is the Fourier transform of $g(r) - 1$, that is,

$$S(k) - 1 = \rho \int [g(r) - 1]e^{i\mathbf{k}\cdot\mathbf{r}}\,d\mathbf{r} = 4\pi\rho \int_0^\infty \frac{\sin kr}{kr}[g(r) - 1]r^2\,dr. \tag{4.19}$$

To perform a numerical evaluation of $g(r)$, we count the average number of particles in the spherical shell between r and $r + dr$ around any given particle in the system and then divide this number by the number of particles in the same spherical shell of the corresponding ideal gas of the same density, $4\pi\rho[(r + dr)^3 - r^3]/3$. If the system is two-dimensional or has a layered structure, we can introduce the two-dimensional radial distribution function, which is useful for characterizing a system trapped on a surface. For a two-dimensional system, the number of particles in the spherical shell is replaced by the number of particles in the ring between r and $r + dr$. For an ideal gas with area density ρ, the number of particles in the ring is $\pi\rho[(r + dr)^2 - r^2]$.

4.3 ^{4}He clusters on a graphite surface

The best known variational quantum Monte Carlo application is by McMillan for the ground state of liquid ^{4}He (McMillan 1965). Over the years, many interesting quantum phenomena have been observed for a collection of helium atoms. We will consider a cluster of ^{4}He atoms on the surface of graphite.

The model Hamiltonian

The Hamiltonian for N ^{4}He atoms on the surface of graphite can be written as

$$H = -\frac{\hbar^2}{2m}\sum_{i=1}^N \nabla_i^2 + \sum_{i=1}^N U_{\text{ext}}(\mathbf{r}_i) + \sum_{i>j=1}^N V(r_{ij}), \tag{4.20}$$

where U_{ext} is the interaction between a helium atom and the graphite surface and V is the interaction between helium atoms. We use a simple, parameterized surface potential introduced by Joly *et al* (1992). The latter potential is given by

$$U_{\text{ext}}(z) = \epsilon U^*(y), \tag{4.21}$$

where

$$U^*(y) = u_0 e^{-\gamma y} - \frac{a_3}{y^3} - \frac{a_4}{y^4}, \tag{4.22}$$

with $y = z/r_m$. The parameters for a helium atom on the surface of graphite are $r_m = 2.9683$ Å, $\epsilon/k = 10.956$ K, $u_0 = 206875.0$, $\gamma = 11.027235$, $a_3 = 6.386763$, and $a_4 = 12.116816$ (Joly $et\ al$ 1992).

The interaction between two helium atoms is taken to be the Aziz potential and is given by (Aziz $et\ al$ 1995)

$$U_{int}(r) = \epsilon V^*(x), \tag{4.23}$$

with

$$V^*(x) = v_0 e^{-\alpha x - \beta x^2} - \left[\frac{c_6}{x^6} + \frac{c_8}{x^8} + \frac{c_{10}}{x^{10}} \right] F(x), \tag{4.24}$$

and

$$F(x) = \begin{cases} e^{-(d/x-1)^2} & x < d \\ 1 & x \geqslant d, \end{cases} \tag{4.25}$$

with $x = r/r_m$. The parameters in equation (4.24) are $v_0 = 186924.404$, $\alpha = 10.5717543$, $\beta = 2.07758779$, $d = 1.438$, $c_6 = 1.35186623$, $c_8 = 0.41495143$, and $c_{10} = 0.17151143$. Note that we have used r_m, the separation distance of the minimum interaction between two helium atoms, to scale all the lengths and ϵ/k to scale all the energies.

We have used certain combinations of constants to simplify the units used in the simulation. For ^{4}He with atomic mass $m = 4.002602$ u, we have $\hbar^2/mk = 12.119232$ K Å^2. Dropping the units in this combination leads to lengths measured in Å and energies in K. The imaginary time $\tau = it/\hbar$ is measured in K^{-1}, which makes τE, with E the energy, dimensionless.

The guide/trial wavefunction

We have constructed a guide/trial wavefunction to have the typical Jastrow form (Mahan 2000) with

$$\Phi(\mathbf{R}) = e^{-J(\mathbf{R})} \prod_{i=1}^{N} \phi(\mathbf{r}_i), \tag{4.26}$$

where $\phi(\mathbf{r}_i)$ is a single-particle wavefunction for the ith particle, and

$$J(\mathbf{R}) = \sum_{i>j} u(r_{ij}) \tag{4.27}$$

is the two-body Jastrow factor. Because we are considering a system of ^{4}He atoms bound to the surface of graphite, all of the single-particle states are bound to the surface in the same manner, for example, a Gaussian function in the direction perpendicular to the surface. Furthermore we imagine that the atoms are spread out along the surface close to a close-packed lattice of lattice constant a_x with L sites, located at $\mathbf{s}_i$. We have

$$\phi(\mathbf{r}_i) = e^{-(z_i-z_e)^2/z_0^2} \sum_{j=1}^{L} e^{-|\mathbf{r}_i-\mathbf{s}_j|^2/r_0^2}, \qquad (4.28)$$

where z_e and z_0 are variational parameters that are on the order of the average distance between the surface and atoms in the cluster, r_0 is a variational parameter that is on the order of a_x. Both a_x and L are variational parameters of the guide/trial wavefunction.

We choose the form of the two-particle function $u(r_{ij})$ as

$$u(r) = \left(\frac{a}{r}\right)^5 + \frac{b^2}{c^2+r^2}, \qquad (4.29)$$

where the first term is from the cusp condition which forces the kinetic energy to cancel the divergent potential energy when the separation between any two particles approaches zero. The second term comes from the phonon contribution in the long-wavelength limit. These parameters can be roughly estimated, for example (Mahan 2000, Glyde 1994), $a \simeq 1.05$ Å and $\sigma \simeq 2.77$ Å.

Symbolically we can write the guide/trial wavefunction as a pure exponential function with

$$\Phi(\mathbf{R}) = e^{-J(\mathbf{R})-X(\mathbf{R})}, \qquad (4.30)$$

where

$$X(\mathbf{R}) = -\sum_{i=1}^{N} \ln \phi(\mathbf{r}_i). \qquad (4.31)$$

The ratio of weights of the distributions from two different configurations in the Metropolis algorithm is

$$\frac{W'}{W} = \frac{|\Phi(\mathbf{R}')|^2}{|\Phi(\mathbf{R})|^2} = e^{-2[J(\mathbf{R}')-J(\mathbf{R})]-2[X(\mathbf{R}')-X(\mathbf{R})]}. \qquad (4.32)$$

We want to choose an initial configuration that is close to the final configuration and also sufficiently flexible so that any number of atoms can be chosen in the simulation. One way to do so is to place the particles in a plane in a close-packed lattice, such as a hexagonal lattice. We first construct lines with particles on each line a distance a_x apart. We can arrange the lines $a_y = a_x \sin \pi/3 = \sqrt{3}a_x/2$ apart, shifting every other line along the x axis back and forth by $a_x/2$. The lattice constant a_x is chosen from the density of a large system and is a little larger than the minimum interaction separation between two atoms, r_m, because of quantum zero-point motion. We do not impose any boundary conditions on the cluster, and a_x is used only to set up the initial configuration and in principle has no bearing on the shape or extent of the cluster when it is in equilibrium.

In this simulation we have chosen a cluster of N atoms that matches the number of sites of the lattice used in the guide/trial wavefunction, that is, $N = L = L_x \times L_y$. This choice allows the program to run with a different number of atoms in the cluster without making a significant change in the program. The distance between the surface and the initial layer of He atoms can be chosen to be the average distance between the surface and the location of the minimum surface potential of a helium atom, $z_e \simeq 2.85$ Å (Carneiro et al 1981). If the cluster is a liquid, we expect its final shape to be close to a disk; otherwise, the particles will form a two-dimensional lattice.

Some preliminary simulation results

We did both variational and diffusion Monte Carlo simulations for ^{4}He clusters on a graphite surface with different values of N. We have searched the parameter space for a reasonable set of parameters in the variational wavefunction to achieve a good start for diffusion Monte Carlo, but have not done an exhaustive search of the parameters because we want to use the wavefunction as a guide for the more powerful diffusion Monte Carlo method. However, a poor guide wavefunction can cause the diffusion Monte Carlo method to converge slowly.

The system that we used to search for a reasonable set of variational parameters in the guide/trial wavefunction had $N = 3 \times 4 = 12$ atoms. The parameters were varied one after another until we reached a point that any further change of the parameters did not produce a noticeable change in the ground-state energy. To remove the dependence on the initial configuration, 15 000 initial Metropolis updates were made before taking any data for averaging. To avoid any significant correlation among the data, 15 additional Metropolis moves were made before taking another value of data for the average. A total of 50 000 energy values were gathered for calculating the average. The data were sorted in two different ways, treating each value or the average of 100 values independently, to ensure the accuracy of the error estimate. The two estimated errors should be about the same if the data are not correlated.

The parameters found using the variational Monte Carlo method are taken to be $a_x = 4.257601$ Å, $z_0 = 0.521$ Å, $r_0 = 15$ Å, $a = 2.770844$ Å, and $b = c = 0$. Here a_x is based on a known commensurate two-dimensional $\sqrt{3} \times \sqrt{3}$ lattice of density (Greywall 1993) $\rho = 0.0636$ Å^{-2}; z_e is taken as the minimum surface potential position (Joly et al 1992), which equals the value found from the variational calculation, a is taken from Giorgini et al (1996), and z_0 and r_0 were found from the variational calculation.

To keep the acceptance probability of the Metropolis moves in the variational Monte Carlo calculations at about 50%, $h_x = h_y = h_z = 0.35$ Å was used for $N = 12$ and a smaller value was used for larger N. The variational ground-state energy per particle obtained from our simulation is $E_0/12 \leqslant -140.40(4)$ K, which decreases slightly with increasing N. A good comparison is the single-particle ground-state energy for this surface potential, -142.2K (Joly *et al* 1992).

We then calculated the ground-state energy of each of the clusters with diffusion Monte Carlo. As expected, the major contribution to the energy comes from the confinement potential of the graphite surface, the same as in the variational Monte Carlo calculations. The parameter $\kappa = 0.1$ K was used in the simulation to achieve the desired ensemble population of 300. We used 2500 data points, each of which contained a weighted-average energy of equation (4.13) from about 300 configurations in an ensemble, taken $15\Delta\tau$ apart, and computed the error for 2500 data points as well as for 50 data points each averaged over 50 of the original data points. Three to nine independent runs were done for each value of $\Delta\tau$. Each run was started from a different initial configuration (with each particle randomly displaced from its lattice site). We summarize our simulation results for the ground-state energies of $N = 12$ and $N = 6 \times 7 = 42$ atoms with different time steps in table 4.1. The $\Delta\tau = 0$ results are obtained from the extrapolation of the simulation data using

$$E_0(\Delta\tau)/N = E_0(0)/N + \beta_1\Delta\tau + \beta_2\Delta\tau^2 + \beta_3\Delta\tau^3 + \cdots, \qquad (4.33)$$

where $E_0(0)$ and β_i can be found from $E_0(\Delta\tau)$, the simulated values for different values of $\Delta\tau$.

We have to ensure that the system has evolved long enough to be in the ground state before any data are taken. The data shown in table 4.1 are taken after 15 000 diffusion time steps.

From table 4.1 we see that each cluster is in a bound state with the ground-state energy per particle lower than the non-interacting single particle energy. We also see that the ground-state energy per particle decreases slightly with increasing N.

We can make a better variational Monte Carlo calculation by a more careful, systematic search of the parameters in the variational wavefunction, or by introducing more terms in the variational wavefunction to improve the representation of the variational wavefunction to the true ground state of the system. The time-step dependence is clearly noticeable. The simulations exhibit significantly fewer fluctuations from independent runs and within each run for $\Delta\tau \leqslant 0.001$ K^{-1}, as indicated by the much smaller statistical variance.

We have tested the program with cluster sizes ranging from $N = 12$ to $N = 13 \times 15 = 195$. Although there is a small cluster size dependence on the energy per particle, the increase of the binding energy (the difference between the single-particle ground-state energy and the energy per particle in the cluster) is small (Sarsa *et al* 2003). The binding energy per particle is less than 0.3 K for $N = 12$, and

Table 4.1. The ground-state energies calculated with different time steps and the corresponding extrapolated values with $N = 12$ and $N = 42$. Each time step is given in $1/(1000$ K$)$ and energy is given in K. The variances from the simulation with the shortest time step are assumed for the extrapolated values.

$\Delta\tau$	2	1	0.5	0.25	0
$E_0/12$	$-145.5(2)$	$-143.12(2)$	$-142.81(1)$	$-142.67(2)$	$-142.46(2)$
$E_0/42$	$-147.4(1)$	$-143.55(2)$	$-143.01(2)$	$-142.90(1)$	$-142.82(1)$

does not go beyond twice this value for a very large cluster. Thus a droplet of ^{4}He formed on the surface of graphite can break up unless the temperature is sufficiently low, for example, much less than 1K.

The behavior of the radial distribution function $g(r)$ yields information about whether the clusters are in a liquid or a solid phase. The computed $g(r)$ for a three-dimensional cluster with $N = 11 \times 13 = 143$ is shown in figure 4.1. Because the particles are not contained in a fixed volume, the number density of the system is not well-defined. We have used the bulk density $\rho = 0.024494$ Å^{-3} to scale $g(r)$ spread out uniformly with the nearest-neighbor distance roughly the same everywhere. If the cluster were a solid, the peaks in $g(r)$ would be of roughly equal height and spaced according to its lattice structure. Hence, our simulation results show that the clusters are likely in a liquid state.

We also considered the projection of the particle coordinates onto a two-dimensional surface. In figure 4.2 we show the two-dimensional radial distribution function taken from another simulation with the same cluster size of $N = 143$. The features of $g_{2D}(r)$ are similar to those in $g_{3D}(r)$, indicating that the positions of the particles are close to a plane near the minimum of the surface potential. The values of $g_{2D}(r)$ are also relative because they are scaled with a uniform density, 0.0636 Å^{-2}.

We have described the diffusion Monte Carlo and associated variational Monte Carlo methods. These methods were demonstrated for the example of ^{4}He clusters trapped on the surface of graphite. Readers can download the sample program used to create the results we have discussed (Pang 2014).

The algorithm that we have implemented in the program used to do the calculations we have discussed involves diffusion, drifting, and branching processes. From the data presented in table 4.1 with different values of $\Delta\tau$, we see that a time

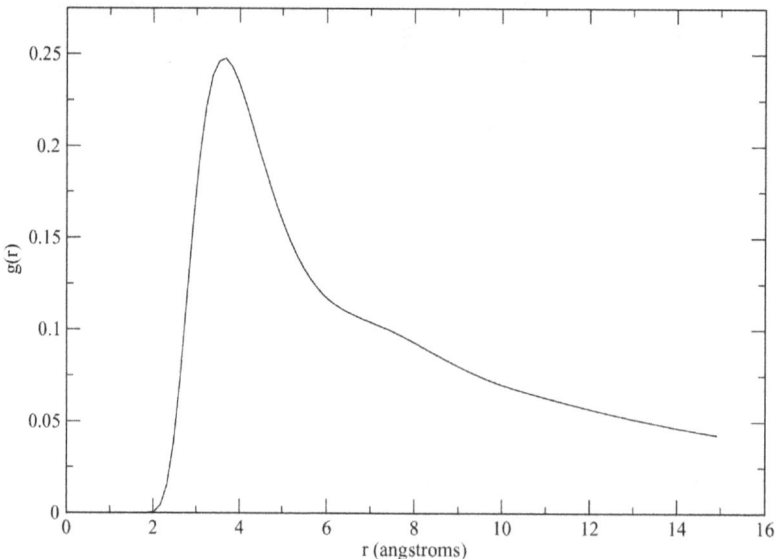

Figure 4.1. The computed three-dimensional radial distribution function for $N = 143$.

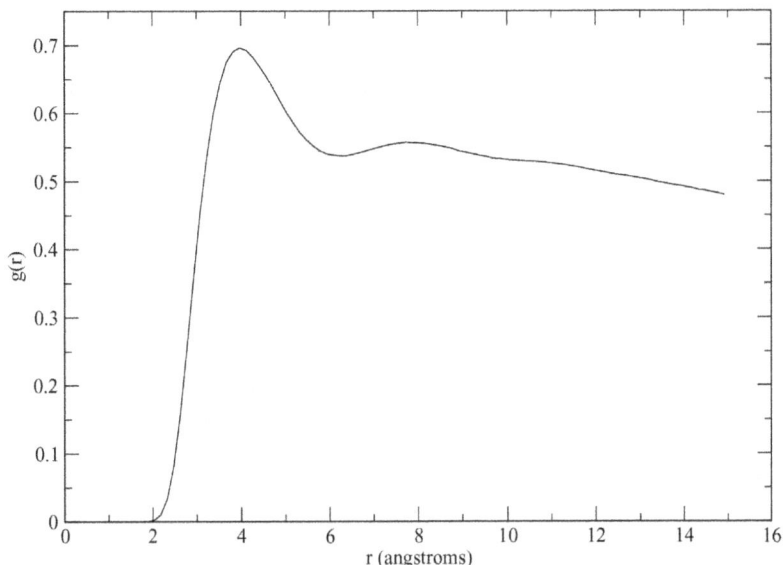

Figure 4.2. The two-dimensional radial distribution function for $N = 143$.

step of $\Delta\tau = 0.0005$ K^{-1} yields meaningful data. The time step is chosen so that the binding energy per particle is accurate to within 0.01 K. If we want to use the code to do calculations for clusters with larger N, or find the binding energy per particle with an error on the order of 0.01 K, the time step should be $\Delta\tau \leq 0.0005$ K^{-1}. A smaller time step would yield higher accuracy, but would require longer runs to reach the ground state.

If we wish to reduce the time-step dependence in the simulation, we can improve the algorithm with a higher-order correction, such as the iteration scheme for the drift step discussed earlier. This higher-order correction scheme requires two additional evaluations of the drift velocity for each time step, but makes the algorithm accurate up to $\Delta\tau^2$ instead of $\Delta\tau$ in the simplest algorithm. Another way to reduce the time-step dependence is to enforce detailed balance at each time step via the Metropolis algorithm, and replace the local energy in the birth/death rate by an average of those of the old and new configurations, and estimate an effective diffusion time step for branching. This estimate requires an additional evaluation of the drift velocity and tracking the attempted moves and accepted moves separately. Both the high-order correction and detailed balance enforcement are implemented in the code.

Exercises

4.1 Find the ground-state energy of a three-dimensional harmonic oscillator with the Hamiltonian $H = -(\hbar^2/2m)\nabla^2 + kr^2/2$, using the diffusion Monte Carlo method with the trial/guide wavefunction $\phi(r) = e^{-r^2/2}$ and natural units of $m = k = \hbar = 1$. Compare your simulation results with the exact result, $E_0 = 3/2$.

4.2 Find the ground-state energy of the hydrogen molecule using the diffusion Monte Carlo method. Assume that the protons are stationary with a fixed separation of 0.741 Å. A possible choice of the trial/guide wavefunction of the two electrons is $\Phi(\mathbf{r}_1, \mathbf{r}_2) = \phi(\mathbf{r}_1)\phi(\mathbf{r}_2)e^{r_{12}/(2a_0)}$, with $\phi(\mathbf{r})$ a linear combination of the 1s atomic orbitals from the two protons and a_0 is the Bohr radius. Show that the cusp condition is satisfied between the two electrons and between an electron and a proton. Improve the trial wavefunction with variational parameters or variations of the functional form while maintaining the cusp condition. Compare the calculated result with the observed dissociation energy of the molecule, 4.52 eV.

4.3 Estimate the ground-state energy of the lithium atom using the diffusion Monte Carlo method. One choice of the trial/guide wavefunction is to combine the 1s and 2s atomic orbitals and a Jastrow factor. Use a Slater determinant for the two electrons with the same spin and reject any time step that causes the system to cross a node of the wavefunction. Compare the calculated result with the observed ground state energy of the atom, -203.4860 eV.

4.4 Calculate the ground-state energy per atom and pair-distribution function of bulk ^{4}He using the diffusion Monte Carlo method. Enforce periodic boundary conditions on the system by moving a particle back to the simulation box on the opposite side when a move takes it to the outside of the box. Use the Aziz potential given in Aziz *et al* (1995) for the interaction between the atoms, $\rho = 0.024494$ Å^{-3} for the density, and $N = 32$ atoms in the simulation box with the starting positions on a face-centered cubic lattice.

4.5 Simulate a system of hard-sphere bosons in a three-dimensional harmonic potential well $V(r) = kr^2/2$ via the diffusion Monte Carlo method. The model can be viewed as a simple model of cold atoms in a trap. Use $\Phi(\mathbf{R}) = \prod_{i=1}^{N} e^{-r_i^2}$ as the trial/guide and reject any move that would cause the separation of any two particles r_{ij} to be less than d, where d is the diameter of the hard spheres. Calculate E_0/N for various values of N with natural units of $m = k = \hbar = 1$. Check your result in the limit $d \to 0$ against $E_0/N = 3/2$. Start the simulation with N on the order of 10 and increase it to 1000. What is the N dependence of the computation time if the same accuracy of calculation is sought?

4.6 Study the electronic structure of the helium atom with the diffusion Monte Carlo method. Assume that the nucleus is fixed at the origin of the coordinates. The key is to find a good parameterized variational wavefunction with proper cusp conditions built in.

4.7 Find the ground-state energy per particle, the density profile, and the pair correlation function of a ^{4}He cluster. How sensitive are the values to the size of the cluster? Assume that the interaction between any two atoms is given by the Lennard-Jones potential and express the results in terms of the potential parameters. What happens if the system is a ^{3}He cluster?

4.8 Use diffusion Monte Carlo to simulate ^{3}He clusters on a graphite surface with the surface potential and interaction used in this article. Construct a guide/trial wavefunction for which the single-particle wavefunction is a Slater determinant of the Gaussian orbitals from the virtual lattice sites. Reject any move that causes crossing a node of the wavefunction. Do the clusters form quantum liquids?

4.9 Implement the diffusion Monte Carlo algorithm in the study of a hydrogen molecule. The two-proton and two-electron system should be treated as a four-body system. Calculate the ground-state energy of the system and compare the result with the best known calculation.

4.10 The structure of the liquid ^{4}He can be studied with the diffusion Monte Carlo method. Develop a program to calculate the ground-state energy per particle and the pair correlation function with the selected variational wavefunction. Assume that the system is in a cubic simulation cell under the periodic boundary condition and use the density measured at the lowest temperature possible and the state-of-the-art calculation of the interaction potential for the simulation.

4.11 Perform the diffusion Monte Carlo simulation of liquid ^{4}He. Is there any significant improvement in the calculated ground-state energy over that of the variational Monte Carlo calculation?

Bibliography

Anderson J C 2007 *Quantum Monte Carlo: Origins, Development, Applications* (New York: Oxford University Press)

Astrakharchik G E 2004 Quantum Monte Carlo study of ultracold gases *PhD Thesis* University of Trento

Aziz R A, Janzen A R and Moldover M R 1995 *Ab initio* calculations for helium: A standard for transport property measurements *Phys. Rev. Lett.* **74** 1586–9

Carneiro K, Passel J, Thomlinson W and Taub H 1981 Neutron-diffraction study of the solid layers at the liquid–solid boundary in ^{4}He films adsorbed on graphite *Phys. Rev.* B **24** 1170–6

Forbert H A and Chin S A 2001 Fourth-order diffusion Monte Carlo algorithms for solving quantum many-body problems *Phys. Rev.* B **63** 144518

Giorgini S, Boronat J and Casulleras J 1996 Diffusion Monte Carlo study of two-dimensional liquid ^{4}He *Phys. Rev.* B **54** 6099–102

Glyde H R 1994 *Excitations in Liquid and Solid Helium* (London: Clarendon Press) p 41

Greywall D S 1993 Heat capacity and the commensurate-incommensurate transition of ^{4}He adsorbed on graphite *Phys. Rev.* B **44** 309–18

Joly F, Lhuillier C and Brami B 1992 The helium-graphite interaction *Surf. Sci.* **264** 419–22

Mahan G 2000 *Many Particle Physics* 3rd ed (Berlin: Springer) ch 10

McMillan W L 1965 Ground state of liquid He^4 *Phys. Rev.* A **138** 442–51

Pang T 2014 Diffusion Monte Carlo: a powerful tool for studying quantum many-body systems *Am. J. Phys.* **82** 980–8

Reynolds P J, Ceperley D M, Alder B J and Lester W A Jr 1982 Fixed-node quantum Monte Carlo for molecules *J. Chem. Phys.* **77** 5593–603

Sarsa A, Mur-Petit J, Polls A and Navarro J 2003 Two-dimensional clusters of liquid ^{4}He *Phys. Rev.* B **68** 224514

Troyer M and Wiese U-J 2005 Computational complexity and fundamental limitations to fermionic quantum Monte Carlo simulations *Phys. Rev. Lett.* **94** 170201

Chapter 5

Path-integral Monte Carlo

The grand challenge in many-body problems is to find a proper solution of the Hamiltonian that contains nontrivial interactions (Mattis 1993). Historically solutions of such problems have yielded great impact to the entire field, for example, the Bethe (1931) ansatz solution of the spin-1/2 one-dimensional Heisenberg model, the Bardeen–Cooper–Schrieffer (BCS) (Bardeen *et al* 1957) solution of conventional superconductivity, and the Laughlin (1983) solution of the fractional quantum Hall effect. These kinds of ingenious solutions appear rarely in physics and only come from the greatest minds of the time. What we do commonly is to develop knowledge of the many-body system gradually and incrementally. This still requires us to probe into the many-body Hamiltonian and try to get a piece of useful information out of the system, which happens to be meaningful, if we are lucky enough.

5.1 Introduction

The original many-body problem is to solve the Schrödinger equation

$$i\hbar\frac{\partial\Psi(\mathbf{R}, t)}{\partial t} = H\Psi(\mathbf{R}, t), \tag{5.1}$$

for a given many-body Hamiltonian H. Here we use $\mathbf{R} = (\mathbf{r}_1, \mathbf{r}_2, ..., \mathbf{r}_N)$ to represent the positions of all the N particles in the system. If H is time-independent, the part of time-dependence can be separated from the rest in the wavefunction

$$\Psi_n(\mathbf{R}, t) = e^{-iE_nt/\hbar}\Psi_n(\mathbf{R}). \tag{5.2}$$

and we end up with the corresponding time-independent Schrödinger equation

$$H\Psi_n(\mathbf{R}) = E_n\Psi_n(\mathbf{R}), \tag{5.3}$$

where E_n is the eigenenergy of the state labeled by quantum number n. The Hamiltonian H can formally be written as the sum of the kinetic energy term

doi:10.1088/978-1-6817-4109-3ch5

$$K = -\frac{\hbar^2}{2m} \sum_{i=1}^{N} \nabla_i^2 \qquad (5.4)$$

and potential energy term

$$U = \sum_{i=1}^{N} U_{ext}(\mathbf{r}_i) + \sum_{i>j=1}^{N} U_{int}(r_{ij}); \qquad (5.5)$$

that is, $H = K + U$. We have made several assumptions in the above expressions: the particles are identical with the same mass m, the external potential energy U_{ext} on each particle is a function of its position, and the interactions between particles U_{int} are pairwise.

Solving the above Schrödinger equation with a large N in general is a formidable task. Instead, we ask what are the properties of the system that can be extracted in order to understand the system. At the very least we are interested in the properties of the ground state and low-lying excited states, including their energies and other expectation values defined by

$$A_n = \frac{\langle \Psi_n | A | \Psi_n \rangle}{\langle \Psi_n | \Psi_n \rangle}, \qquad (5.6)$$

where A is the corresponding physical operator.

Rarely an analytical solution of a many-body problem can be obtained. Approximate or numerical solutions are the best alternatives in most cases. There are several numerical methods that have been used to probe the ground-state and low-lying excited state properties of many-body systems and they scale with the size of the system quite differently. For example, the direct diagonalization schemes, such as the configuration-interaction method (Cramer 2002), scale with the size of the system as N^6 while the mean-field approaches such as the Hartree–Fock method (Slater 1974) and density-functional theory (Hohenberg and Kohn 1964) with a local approximation (Kohn and Sham 1965) of the exchange–correlation interactions scale with the size of the system as N^3. That is why exact diagonalization is typically used for systems with a small N and mean-field method is more meaningful for systems with large N.

Quantum Monte Carlo methods typically scale with the size of the system as N^4 and are appropriate to systems with intermediate N. We have done a tutorial on the variational and diffusion quantum Monte Carlo simulation methods (Pang 2014). In this chapter, our focus will be on the temperature-dependent properties of the system and one of the most powerful tools in probing the finite-temperature properties of a many-body system is the path-integral Monte Carlo simulation.

5.2 The propagation of a quantum state

When we study the dynamics of a particle of mass m in the one-dimensional quantum world, its behavior is described by the time-dependent Schrödinger equation

$$\left(H - i\hbar\frac{\partial}{\partial t}\right)\psi(x, t) = 0, \tag{5.7}$$

where $\psi(x, t)$ is the wavefunction of the particle at position x and time t,

$$H = -\frac{\hbar^2}{2m}\frac{\partial^2}{\partial^2 x} + U_{ext}(x)$$

is the Hamiltonian of the particle, and $\hbar$ is the Planck constant. The first term in H is the kinetic energy term and $U_{ext}(x)$ is the external potential energy of the particle at x.

If the wavefunction at a given location x_0 and a given time t_0 is known, the solution of this partial differential equation at arbitrary location x and time t can formally be written as

$$\psi(x, t) = \int G(x, t; x_0, t_0)\psi(x_0, t_0)dx, \tag{5.8}$$

where $G(x, t; x_0, t_0)$ is a propagator, known as Green's function, of the particle from position x_0 and time t_0 to position x and time t, and is the solution of the inhomogeneous equation

$$\left(H - i\hbar\frac{\partial}{\partial t}\right)G(x, t; x_0, t_0) = -\delta(x - x_0)\delta(t - t_0).$$

Note that we have used the Dirac delta functions on the right of the above equation. We can write Green's function in terms of the eigenstates of the time-independent Schrödinger equation:

$$G(x, t; x_0, t_0) = \sum_n \phi_n(x)\phi_n^*(x_0)e^{-iE_n\Delta t/\hbar}, \tag{5.9}$$

where $\Delta t = t - t_0$. Wavefunction $\phi_n(x)$ is the nth eigenstate of the eigen equation

$$\left(H - E_n\right)\phi_n(x) = 0,$$

where E_n is the corresponding eigenenergy of the particle. For a more detailed exposition of Green's function outlined here, see Economou (2006).

Now considering the trivial case of a constant potential $U_{ext}(x) = U_0$, the solution of the time-independent Schrödinger equation is given by plane waves, that is,

$$\phi_k(x) = \frac{1}{\sqrt{2\pi}}e^{ikx},$$

with corresponding energy

$$E_k = \frac{\hbar^2 k^2}{2m} + U_0,$$

where k is the wave number that labels the eigenstate $\phi_k(x)$. The analytic form of Green's function can then be found from carrying out the integration:

$$G(x, t; x_0, t_0) = \frac{1}{2\pi} \int e^{-iE_k \Delta t/\hbar + ik\Delta x} dk = \sqrt{\frac{m}{2\pi i\hbar\Delta t}} e^{i\frac{\Delta t}{\hbar}\left(\frac{m}{2}v^2 - U_0\right)}, \quad (5.10)$$

where $v = (x - x_0)/\Delta t$ can be interpreted as the drift velocity on the particle. Note that this analytic expression is possible because of a constant external potential $U_{ext}(x) = U_0$.

Then what happens if $U_{ext}(x)$ changes with x? First we can examine the situation with $\Delta t \to 0$, that is, the propagation of the state over an infinitesimal elapsed time. Under such a condition we can replace U_0 by any value of $U_{ext}(x)$ between x_0 and x. In rare cases, this fails when $U_{ext}(x)$ is a fast changing function of x.

Furthermore Green's function is convolutionary from its nature of being a propagator, that is,

$$G(x, t; x_0, t_0) = \int G(x, t; x', t')G(x', t'; x_0, t_0)dx',$$

where x' can be interpreted as an intermediate step in the propagation from time t_0 to time t'. Intuitively we can view that the particle is going through a trajectory or path starting from x_0 at t_0 and reaching x at t. However, the intermediate position x' is integrated over the entire space. This means that this integral covers all the possible paths from x_0 at t_0 to x at t because x' takes all the possible values via integration.

Combining the two elements of Green's function discussed above, we can decompose the propagation over a finite elapsed time to a sum of infinite number of steps, each of which goes over an infinitesimal time interval $\tau = t_j - t_{j-1} = \Delta t/M$ with $M \to \infty$. Green's function then becomes

$$G(x, t; x_0, t_0) = \lim_{M\to\infty} \int G(x_M, t_M; x_{M-1}, t_{M-1}) \prod_{j=1}^{M-1} G(x_j, t_j; x_{j-1}, t_{j-1}) \, dx_j,$$

with $x_M = x$ and $t_M = t$. The propagation over each time interval is then given by

$$G(x_j, t_j; x_{j-1}, t_{j-1}) \simeq \sqrt{\frac{m}{2\pi i\hbar\tau}} e^{i\frac{\tau}{\hbar}\left(\frac{m}{2}v_j^2 - U_j\right)},$$

where $v_j = (x_j - x_{j-1})/\tau$. The approximation in the above propagator comes from treating the potential between x_j and x_{j-1} as a constant, $U_j = U(x_j)$. The integral representation of $G(x, t; x_0, t_0)$ here is known as Feynman's path integral (Feynman and Hibbs 1965), which provides a formulation to carry out Green's function order by order on τ. The reason to call it a path integral is because the silent variables that are integrated out, $x_1, x_2, ..., x_{M-1}$, appear to be points on a flexible path that stretches everywhere in the entire space; see an illustration of different paths between

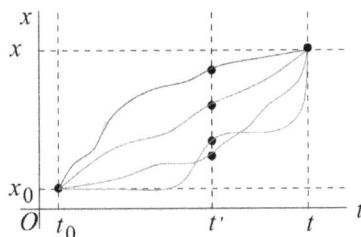

Figure 5.1. An illustration of different paths between two points in space and time in the path-integral formulation of the propagator with the points on the paths corresponding to an intermediate time t' marked.

the two points of the propagator in figure 5.1. We can collect all the exponents of all the propagators in a sum and thus rewrite the integral as

$$G(x, t; x_0, t_0) = \lim_{M \to \infty} \left(\frac{m}{2\pi i \hbar \tau} \right)^{M/2} \int e^{i\frac{\tau}{\hbar} \sum_{j=1}^{M} \left(\frac{m}{2} v_j^2 - U_j \right)} \prod_{j=1}^{M-1} dx_j. \tag{5.11}$$

This expression is what can be used in a simulation with a large M and can take advantage of the Metropolis algorithm for an integral with a large number of variables.

5.3 Single-particle system

A straightforward interpretation of quantum mechanics is that $p_n(x; 0) \propto |\phi_n(x)|^2$ represents the spatial distribution of a particle on the state labeled by n. A Boltzmann factor comes in the distribution when we include the probability of occupying a state at finite temperature; that is,

$$p_n(x; \beta) = \frac{1}{Z} p_n(x; 0) e^{-\beta E_n}, \tag{5.12}$$

where $\beta = \frac{1}{kT}$ is the inverse temperature and Z is the normalization factor known as the partition function, given by,

$$Z = \sum_n \int p_n(x; \beta) dx = \mathrm{Tr} \, \rho(x, x_0; \beta), \tag{5.13}$$

where

$$\rho(x, x_0; \beta) = \sum_n \phi_n(x, \beta) \phi_n^*(x_0) e^{-\beta E_n} \tag{5.14}$$

is called the density matrix of the particle. The trace operation Tr integrates the diagonal elements of the density matrix; that is,

$$\mathrm{Tr} \, \rho(x, x_0; \beta) = \int \rho(x, x; \beta) dx.$$

The average of a physical quantity described by operator A is given by

$$\langle A \rangle = \sum_n \int A_n(x) p_n(x; \beta) \mathrm{d}x = \frac{1}{Z} \operatorname{Tr} A\rho(x, x_0; \beta), \qquad (5.15)$$

where $A_n(x)$ is the function form of operator A when operating on state $\phi_n(x)$.

The mathematical structure of the density matrix is the same as that of Green's function if we identify β with $\mathrm{i}\Delta t/\hbar$. Following the argument for the path-integral representation of Green's function of the preceding section, we find the path-integral representation for the density matrix:

$$\rho(x, x_0; \beta) = \lim_{M \to \infty} \left(\frac{m}{2\pi\hbar\tau} \right)^{M/2} \int \mathrm{E}^{-\frac{\tau}{\hbar} \sum_{j=1}^{M} \left(\frac{m}{2} v_j^2 + U_j \right)} \prod_{j=1}^{M-1} \mathrm{d}x_j, \qquad (5.16)$$

where $\tau = \hbar\beta/M$, $v_j = (x_j - x_{j-1})/\tau$, and $x = x_M$. Note that the exponent is real now and the two terms in the exponent carry the same sign.

Then we can evaluate any physical quantity of the particle at a given temperature through the path-integral representation. For example, the average energy of the particle is given by

$$E = -\frac{\partial \ln Z}{\partial \beta} = \frac{\hbar}{2\tau} - \frac{1}{M} \left\langle \sum_{j=1}^{M} \left(\frac{m}{2} v_j^2 - U_j \right) \right\rangle, \qquad (5.17)$$

where the average is defined as

$$\langle A \rangle = \frac{1}{Z} \int A(x_1, x_2, \ldots, x_M) W(x_1, x_2, \ldots, x_M) \prod_{j=1}^{M} \mathrm{d}x_j, \qquad (5.18)$$

with the weight

$$W(x_1, x_2, \ldots, x_M) = \left(\frac{m}{2\pi\hbar\tau} \right)^{M/2} \mathrm{e}^{-\frac{\tau}{\hbar} \sum_{j=1}^{M} \left(\frac{m}{2} v_j^2 + U_j \right)}, \qquad (5.19)$$

with $x_0 = x_M$. The partition function itself is given by

$$Z = \int W(x_1, x_2, \ldots, x_M) \prod_{j=1}^{M} \mathrm{d}x_j. \qquad (5.20)$$

Note that all the positions are integrated out in the partition function. When Green's function is represented in the path-integral form, the propagator along the path can be visualized as a flexible chain that starts at x_0 and t_0 and ends at x and t. However, the paths of the integrals for the partition function and the corresponding averages here are in the shape of flexible rings because $x_0 = x_M$. This is an important feature to look after when quantum statistics is enforced for a many-body system and will be discussed later. An illustration of such a path in a three-dimensional space is given in figure 5.2. The Metropolis algorithm is most suitable to carry out the integrals here.

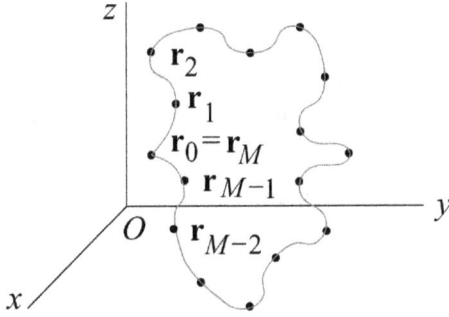

Figure 5.2. An illustration of a closed path in a three-dimensional space in the integration of the partition function.

Formally we can split the average energy into the average kinetic energy

$$\bar{K} = -\frac{m}{\beta Z}\frac{\partial Z}{\partial m} = \langle K \rangle,$$

with

$$K = \frac{\hbar}{2\tau} - \frac{1}{M}\sum_{j=1}^{M}\frac{m}{2}v_j^2,$$

and the average potential energy

$$\bar{U} = E - \bar{K} = \langle U \rangle,$$

with

$$U = \frac{1}{M}\sum_{j=1}^{M}U_j.$$

We can also evaluate all sorts of thermal coefficients of the system in the same manner through partial derivatives of the partition function. For example, the heat capacity of the the system can be calculated from

$$C = \frac{\partial E}{\partial T} = -k\beta^2\frac{\partial E}{\partial \beta} = k\left\langle \beta^2(K + U)^2 + 2\beta U \right\rangle - k\beta^2\langle K + U \rangle^2 - \frac{Mk}{2}. \quad (5.21)$$

Different forms of the average kinetic energy found through integration by parts have been used in calculations either to stabilize the simulation or to check the convergence of a different calculation.

In interpreting the exponent of the density matrix, we have used $v_j = (x_j - x_{j-1})/\tau$, implying a velocity equivalent. Thus the term $mv_j^2/2$ looks like a term of classical kinetic energy, or $mv_j^2/2 + U_j$ looks like the total energy of a particle on a chain of M particles. But if the path is literally visualized as a polymer, or a chain of M beads, the term $mv_j^2/2$ can also be written as $\kappa(x_j - x_{j-1})^2/2$ and viewed as a term of potential energy on the spring that connects the two adjacent beads with spring constant $\kappa = m/\tau^2$. Either picture can help us visualize the quantum particle behaving like a classic polymer. Or in other words, there is an equivalence between a quantum object of of d dimensions ($d = 0$ for a point particle and 1 for a string/polymer, for example) and a classical object of $d + 1$ dimensions.

5.4 Quantum many-body systems

Now we can extend what we have worked out for a single particle to a many-body system with the partition function

$$Z = \mathrm{Tr}\,\rho(\mathbf{R}, \mathbf{R}_0; \beta)$$

where $\mathbf{R} = (\mathbf{r}_1, \mathbf{r}_2, ..., \mathbf{r}_N)$, is the position vector of N particles. Assuming that the space is d-dimensional, with $d = 1, 2, ...$, $\mathbf{R}$ contains dN coordinates. Replacing x_j with $\mathbf{R}_j$ in the weight, partition function, and all the averages, we reach the corresponding expressions for the many-body systems. For example, the weight function now becomes

$$W(\mathbf{R}_1, \mathbf{R}_2, ..., \mathbf{R}_M) = \left(\frac{m}{2\pi\hbar\tau}\right)^{dNM/2} e^{-\frac{\tau}{\hbar}\sum_{j=1}^{M}\left(\frac{m}{2}\mathbf{v}_j^2 + U_j\right)}, \tag{5.22}$$

with $\mathbf{R}_j$ for $j = 0, 1, ..., M$ being a point on the path, constrained by $\mathbf{R}_0 = \mathbf{R}_M$. Here the velocity

$$\mathbf{V}_j = \frac{\mathbf{R}_j - \mathbf{R}_{j-1}}{\tau}$$

and the potential energy

$$U_j = U_{\mathrm{ext}}(\mathbf{R}_j) + U_{\mathrm{int}}(\mathbf{R}_j),$$

with U_{ext} being the external potential energy of all the particles and U_{int} being the interaction between all the particles, which is pairwise in most cases.

The statistical average of any physical quantity is given by

$$\langle A \rangle = \frac{1}{Z} \int A(\mathbf{R}_1, \mathbf{R}_2, ..., \mathbf{R}_M) W(\mathbf{R}_1, \mathbf{R}_2, ..., \mathbf{R}_M) \prod_{j=1}^{M} d\mathbf{R}_j. \tag{5.23}$$

For example, the average kinetic energy can be found from $\bar{K} = \langle K \rangle$, where

$$K = \frac{dN\hbar}{2\tau} - \frac{1}{M}\sum_{j=1}^{M}\frac{m}{2}\mathbf{v}_j^2,$$

and the corresponding average potential energy can be found from $\bar{U} = \langle U \rangle$, where

$$U = \frac{1}{M}\sum_{j=1}^{M} U_j,$$

with U_j given at each point $\mathbf{R}_j$ along the path.

However, in addition to the interaction, there is a major difference between a many-body quantum system and the corresponding single-particle system because of quantum statistics. For example, the many-body wave function stays the same when we interchange two particles in a Bose system but changes its sign in a Fermi system. This is reflected in the density matrix as well. So the correct density matrix must reflect this symmetry:

$$\rho_\pm(\mathbf{R}, \mathbf{R}_0; \beta) = \frac{1}{N!} \sum_P (\pm)^P \rho(\mathbf{R}, \mathbf{PR}_0; \beta),$$

where + is for a Bose system and − is for a Fermi system and P represents a permutation operation.

The permutation operations for a Bose system do not cause any problem because the weight is still positive so that the Metropolis algorithm can still be applied; the only complication is that permutations must be made in the simulations to accommodate the many-body behavior of the system. However, a serious problem appears for a Fermi system because of the sign change in front of each permutated contribution in the density matrix, known as the *fermion sign problem*. At the moment, different strategies have been proposed to deal with different situations, but none has been presented as a true solution. It turns out that the fermion sign problem is NP hard and may never be solved truly.

5.5 Physical properties of extended systems

Modern computers have reached the point that a significantly large quantum many-body system can be simulated to represent the experimental system well. For a bulk system or a large cluster of quantum particles, this is important. For example, the pressure of a many-body quantum system can be calculated from the virial theorem:

$$P = \frac{2}{3V}\left(\bar{K} - \Phi\right) \tag{5.24}$$

where Φ is the average

$$\Phi = i\left\langle r_{jk} \frac{\partial U_{\text{int}}(r_{jk})}{\partial r_{jk}} \right\rangle.$$

We have assumed that the external potential $U_{\text{ext}} = 0$ and the interaction potential is pairwise,

$$U_{\text{int}}(\mathbf{R}) = \sum_{j>k=1}^{N} U_{\text{int}}(r_{jk}).$$

Of course, we must also deal with the finite-size effect when a bulk system is modeled in a simulation. Normally we set up a simulation box and impose periodic boundary conditions during the simulation with particles going out of the simulation box coming back from the other side of the box. If the interaction is short-ranged we can truncate it within the simulation box. Long-range interactions such as the Coulomb force will need the inclusion of the interactions between the simulation box and its images, such as the Edward summation.

We can also calculate the radial distribution function of the system by counting. Note that when the simulation is along the entire path, only the particles from different rings at the same time slice are counted. In other words, the radial distribution is given by

$$g(r) = \frac{1}{MN\rho} \left\langle \sum_{j>k=1}^{N} \sum_{l=1}^{M} \delta\left(r - \left|\mathbf{r}_{j,l} - \mathbf{r}_{k,l}\right|\right) \right\rangle,$$ (5.25)

where $l = 1, 2, ..., M$ represents different time slices along the closed path (ring).

5.6 Cold atoms

About 90 years ago, Bose (1924) and Einstein (1925) predicted a phenomenon of having the majority of particles accumulating on the single-particle ground state if the particles obey the Bose statistics and the interaction between any two particles is vanishingly small. This phenomenon known as *Bose–Einstein condensation* requires cooling a system of many bosons down to near a nanokelvin and was therefore considered as a pure intellectual exercise rather than a reality. Over time, experimentalists have tried to create systems to come close to such an ideal case, for example, by making a helium cluster near a superfluid surface.

The situation changed in the mid-1990s with the availability of laser cooling that can drive a system to the nanokelvin range. The remaining task was to keep a cluster of nearly noninteracting particles together. Two techniques were perfected during that period. One uses the potential well created on a magnetic dipole carried by alkali atoms in a magnetic field and the other uses three pairs of lasers to trap and cool a cluster of alkali atoms concurrently. Three experimental groups published their findings in creating a condensate of alkali atoms almost at the same time (Anderson *et al* 1995, Bradley *et al* 1995, Davies *et al* 1995).

Among the large number of theoretical activates after the discovery of Bose–Einstein condensation in alkali atom clusters, we have performed a series of path-integral quantum Monte Carlo simulations of hard-sphere bosons trapped in a harmonic potential to model the cold atoms (Pearson *et al* 1998, for example). For an isotropic three-dimensional system, the external potential is

$$U_{\text{ext}}(\mathbf{r}_i) = \frac{m\omega^2}{2} r_i^2,$$ (5.26)

and the hard-sphere assumption is represented by the interaction between the ith and jth particles in the system as

$$U_{\text{int}}(r_{ij}) = \begin{cases} \infty & \text{for } r_{ij} < a, \\ 0 & \text{otherwise,} \end{cases}$$ (5.27)

where a is the diameter of the hard spheres. We have sampled systems up to 1000 particles.

We have calculated the heat capacity C, and the condensation fraction N_0/N for systems with number of particles ranging from 10 to 1000. As discussed earlier in this chapter, the heat capacity of a given boson system can be expressed in terms of averages of the energies and their combinations. The fraction of particles in the ground state, that is, the condensation fraction, N_0/N is estimated from the particle on permutation loops greater than the linear dimension of the systems, measured by $N^{1/3}$. Intuitively this estimation makes sense because the particles on a loop of the path integral are coherent with each other and the loops extend to the entire cluster corresponding to the lowest states. This was numerically shown in a previous path-integral Monte Carlo study.

The units of the physical quantities are set with $\hbar = k = m = \omega = 1$ to simplify the simulations; we thus have $\lambda = 1$, $a = a/\lambda$, and $T = kT/(\hbar\omega)$. In figure 5.3, we show the temperature dependence of the the condensation fraction for a system of 1000 particles with different hard-sphere diameters. The small diameter case with $a = 0.02$ appears to be very similar to the noninteracting case ($a = 0$) and agrees well with the analytical result. When the interaction increases with a larger a, condensation fraction goes down at the same temperature but still remains a sharp slope for us to estimate the transition temperature for each given hard-sphere diameter.

We have also calculated the condensation fractions of systems with 10 and 100 particles and found a similar trend of behavior in both of them. The condensation fraction for systems with small hard-sphere diameters, such as $a = 0.02$, is very close to that of a noninteracting or ideal system. When the diameter of the spheres becomes significant, for example, $a = 0.2$, the change in the critical temperature toward zero kelvin becomes significant.

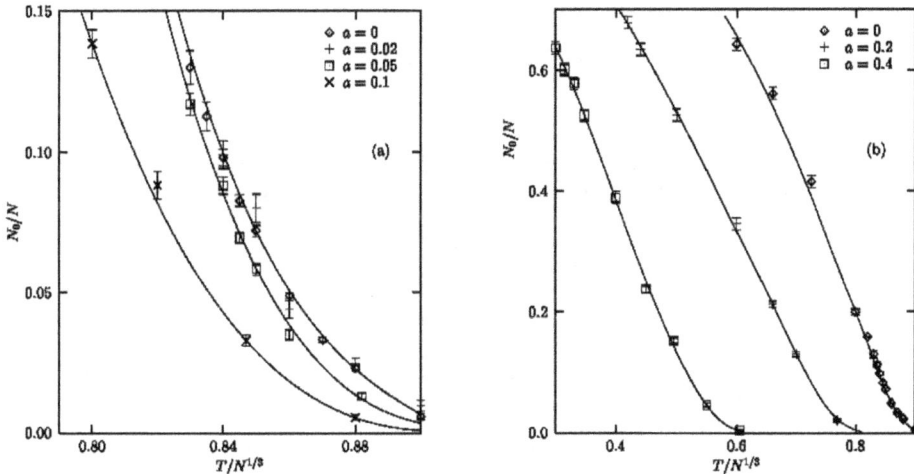

Figure 5.3. The condensation fraction N_0/N of 1000 hard-sphere bosons in an isotropic three-dimensional parabolic potential with $\hbar = m = \omega = 1$ (Pearson *et al* 1998). Simulation results for small and large hard-sphere diameter a are presented in (*a*) and (*b*), respectively. The lines are fitted with the least-squares method.

The specific heat (heat capacity of each particle), C/N for systems with different total number of particles, N, up to $N = 1000$, are evaluated and shown in figure 5.4.

The relative positions of the peaks on the specific heat–temperature curves can provide estimates of the changes in the critical temperature of the system. No significant change is found in the critical temperature when the diameter of the hard spheres is small, for example, $a < 0.05$. The critical temperature changes to a lower value when the diameter of the hard spheres becomes significant, for example, $a > 0.1$. The change in critical temperature becomes more noticeable for larger systems; for example, at $a = 0.05$, the drop in the critical temperature for the system with $N = 1000$ particles is much more significant than that in the system with $N = 10$ particles. For an ideal case with infinite number of particles, a divergence is expected in the specific heat for a second-order transition.

Based on the temperature dependence of the condensation fraction and the specific heat for various system sizes, we can build a curve for diameter dependence of the change in the critical temperature of the system. These data are collectively shown in figure 5.5. When the interaction between the particles is weak; that is, a is small, for example, $aN^{1/6} < 0.2$, no significant change is found in the critical temperature of the system. As shown in the inset in figure 5.5, practically no change in the critical temperature is observed. For systems with stronger interaction, that is, a larger a, the drop in the critical temperature of the system becomes noticeable. The change is almost linear in the region $aN^{1/6} \in [1, 3]$.

We have further expanded our study to a mixture of two hard-sphere boson species with different diameters in an asymmetric harmonic trap (Ma and Pang 2004). The Hamiltonian of the system is

$$H = H_1 + H_2 + U_{\text{int}}, \qquad (5.28)$$

where

$$H_i = -\frac{\hbar^2}{2m_i} \sum_{l=1}^{N_i} \nabla_l^2 + \sum_{l=1}^{N_i} U_{i\text{ext}}(\mathbf{r}_l), \qquad (5.29)$$

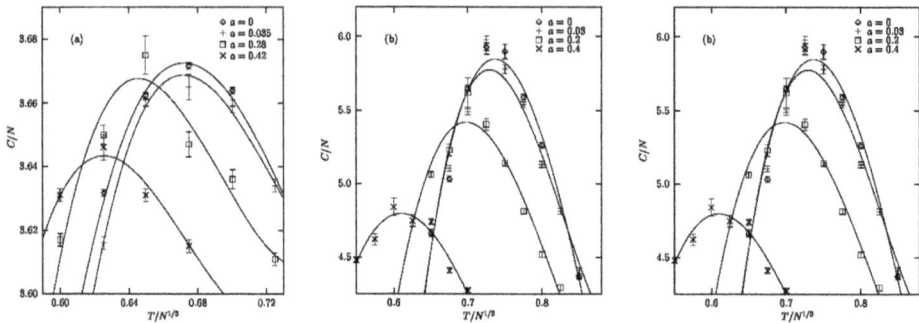

Figure 5.4. The heat capacity per particle C/N of (a) 10, (b) 100, and (c) 1000 hard-sphere bosons in an isotropic three-dimensional parabolic potential with $\hbar = m = \omega = 1$ (Pearson *et al* 1998). We have used the least-squares fit to produce the lines shown.

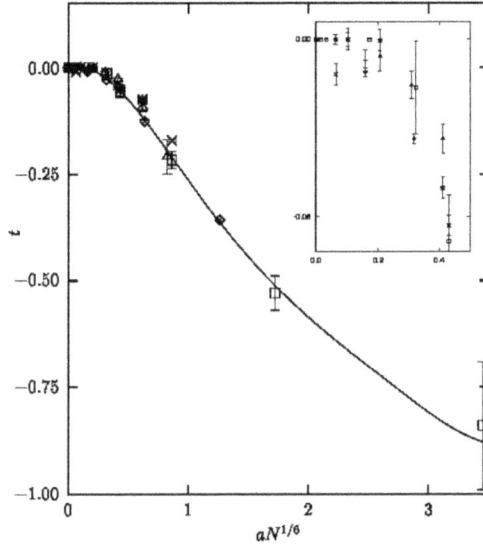

Figure 5.5. The relative change in the critical temperature $t = (T_c - T_c^0)/T_c^0$ in a system of N hard-sphere bosons of various diameters a, trapped in an isotropic three-dimensional harmonic potential with $\hbar = m = \omega = 1$ (Pearson *et al* 1998). The data shown are estimated from the temperature dependence of the simulated condensation fraction N_0/N for $N = 10$ (triangles), $N = 100$ (boxes), and $N = 1000$ (diamonds), and from the temperature dependence of the specific heat C/N with $N = 10$ (stars), $N = 100$ (diagonal crosses), and $N = 1000$ (vertical crosses). A line is drawn to reflect the trend of the data.

with $i = 1,2$ and the external potential

$$U_{i\text{ext}}(\mathbf{r}_l) = \frac{m_i \omega_i^2}{2}\left(x_l^2 + y_l^2 + \lambda^2 z_l^2\right). \tag{5.30}$$

The parameter $\lambda = \omega_z/\omega_\perp$ provides a measure of the aspect ratio between the axial confinement along the z direction and the radial confinement in the xy plane, with $\omega_\perp = \omega_x = \omega_y$. With $\lambda \gg 1$, the system is in the shape of a flat disk, whereas with $\lambda \ll 1$ the system is in a long cigar shape. The limits include $\lambda = 1$ (spherically symmetric), $\lambda = 0$ (confined along the z axis), and $\lambda = \infty$ (confined in the xy plane).

The interaction potential $U_{\text{int}}(t_{jk})$ reflects the hard spheres in each of the two species, modeled after the s-wave scattering lengths of the experimental systems, with a_{11} and a_{22} being the diameters of the particles in species 1 and 2, respectively, and $a_{12} = (a_{11} + a_{22})/2$ being the potential range between two particles from two different species. Different masses m_1 and m_2 and total numbers of particles N_1 and N_2 are used in the simulation, with $N = N_1 + N_2$ being the total number of particles in the two-species system.

We have performed simulations by carefully choosing the total number of time slices and the size of the Monte Carlo steps to ensure a good convergence of the physical quantities calculated, including the density $\rho_i(\mathbf{r})$ and the in-plane angular pair-correlation function

$$\Gamma_i(\varphi) = \langle \delta(\varphi - |\varphi_l - \varphi_k|) \rangle, \tag{5.31}$$

for $l \neq k$. In order to examine the condensate profiles against the planar symmetry of the trap, temperature is set much lower than the critical temperature and all the particles are projected into the xy plane when we take snapshots or calculate the average radial density $\rho_i(\rho)$ and angular correlation function $\Gamma_i(\varphi)$, where $\rho = \sqrt{x^2 + y^2}$ is the in-plane radius and φ is the relative azimuthal angle. The density is normalized by

$$2\pi \int_0^\infty \rho_i(\rho)\rho \, d\rho = N_i. \tag{5.32}$$

In the left panel of figure 5.6, three sets of simulation results for different aspect ratio λ under the same trapping frequency for both species at $T/N^{1/3} = 0.1$ are shown (Ma and Pang 2004). It is clear that the symmetry of the Hamiltonian is preserved in the

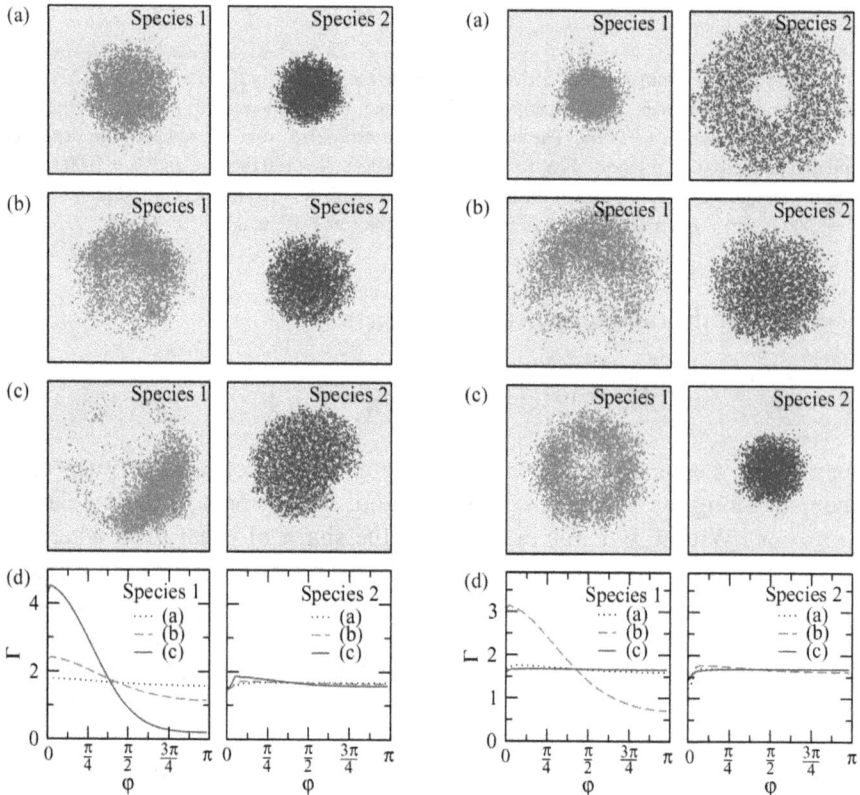

Figure 5.6. Snapshots of multi-exposures of particles viewed along the z direction for each species in a double-condensate mixture (Ma and Pang 2004). The left panel is for $m_2/m_1 = 4$, $N_1 = N_2 = 100$, $a_{11} = 0.2$, and $a_{22} = 0.4$, with different aspect ratio $\lambda = \omega_z/\omega_\perp$: (a) $\lambda = 1$; (b) $\lambda = 4$; and (c) $\lambda = 16$, with the corresponding in-plane angular pair-correlation functions $\Gamma_i(\varphi)$ shown in (d), and the right panel is for $m_2/m_1 = 4$, $N_1 = N_2 = 100$, $\lambda = \omega_z/\omega_\perp = 16$, and $a_{11} = 0.15$ fixed, for different ratio $\eta = a_{11}/a_{22}$: (a) $\eta = 1/3$; (b) $\eta = 1/2$; and (c) $\eta = 1$, with the corresponding in-plane angular pair-correlation functions $\Gamma_i(\varphi)$ shown in (d).

condensate profiles when the trap is spherically symmetric, that is, $\lambda = 1$. For both $\lambda = 4$ and $\lambda = 16$, the profiles are obviously asymmetric with a random orientation due to spontaneous symmetry-breaking. The asymmetry becomes more severe and the correlation at small angle, as shown in part (d) on the left of figure 5.6, becomes much stronger as the planar symmetry in each of the two condensates is broken if $\lambda \neq 1$; the asymmetry becomes more severe when λ is further away from 1, especially for the lighter species.

The symmetry-breaking in one species compensates the other because the center of mass of the system, under the given conditions, remains at the center of the trap, resulting in a stronger deformation in the lighter species. We have done one more simulation with $\lambda < 1$ and not seen any symmetry-breaking of the profiles in the xy plane. This is consistent with the findings of previous mean-field calculations based on the Gross–Pitaevskii equation, which concluded that the symmetry is broken in the direction of the weakest trapping frequency.

We have also examined how the system responds to the change of the ratio $\eta = a_{11}/a_{22}$, which is a measure of the imbalance of the particle sizes between the two species. In the right panel of figure 5.6, we show three sets of snapshots with different η while having other parameters $m_2/m_1 = 4$, $N_1 = N_2 = 100$, $\lambda = 16$, and $a_{11} = 0.15$ fixed. The planar symmetry of the particle distributions is broken first and then restored when the ratio is varied monotonically.

For the case of $\eta = 1/3$, the system shows a condensed core of species 1, surrounded by an outer ring of species 2. Both condensates appear to have the planar symmetry of the trap. When η is increased to 1/2, the system undergoes a quantum phase transition from two separate condensates to a binary mixture with each breaking the planar symmetry. This is evident from both snapshots and the correlation functions shown. The correlation at small angle increases when the symmetry is broken. Note that the condensate of species 1 is expected to be more off-centered (four times) because the center of mass of the whole system must remain at the trap center. One can see from the snapshots that the binary phase is formed from the outward movement of species 1 plus the inward movement of species 2. Therefore, increasing η further can result in exchanging the roles of the two species. This is precisely what happens when η is increased to 1: the system shows a condensed core of species 2, surrounded by an outer ring of species 1. The correlation at small angle decreases and the planar symmetry of the Hamiltonian gets restored in the condensate profiles at the same time.

We have also calculated many other properties of a boson mixture in a trap and interested readers can find details in our publications (Ma and Pang 2004, 2006).

Exercises

5.1 Develop a path-integral Monte Carlo program to study a simple harmonic oscillator with potential energy $U_{ext}(x) = m\omega^2 x^2/2$ for any given mass m and angular frequency ω. Evaluate the heat capacity of the particle at different temperature.

5.2 Simulate N identical, noninteracting bosons, each of mass m, trapped in an isotropic, three-dimensional harmonic potential with $U_{\text{ext}}(r_i) = kr_i^2/2$. Estimate the N dependence of the transition temperature. Does the calculated transition temperature have the right limit at $k \to 0$ and $N \to \infty$ but a uniform density?

5.3 Carry out path-integral quantum Monte Carlo simulation of a hard-sphere boson cluster in an anisotropic harmonic trap with a trapping potential

$$V(\mathbf{r}) = \frac{m\omega^2}{2}\left(\lambda z^2 + x^2 + y^2\right),$$

where m is the mass of a particle and ω and λ are parameters of the trap. Find the condensation temperature of the system for a set of different values of the total number of particles, hard-sphere radius, ω, and λ. What happens if there are more than one species of bosons in the trap?

5.4 Calculate the temperature dependence of the specific heat of bulk ^{4}He using the path-integral Monte Carlo method. Enforce periodic boundary conditions on the system by moving a particle back to the simulation box on the opposite side when a move takes it to the outside of the box. Use the Aziz potential given in Aziz et al (1995) for the interaction between the atoms, $\rho = 0.024494$ Å^{-3} for the density, and $N = 32$ atoms in the simulation box with the starting positions on a face-centered cubic lattice.

5.5 Carry out a path-integral Monte Carlo study of a ^{3}He cluster on a graphite surface. Use the Aziz potential (Aziz et al 1995) for the interaction between two atoms and Joly, Lhuillier, and Brami potential (Joly et al 1992) for the interaction between an atom and the surface. Is the system a superfluid or normal fluid approaching zero temperature?

5.6 Develop a path-integral Monte Carlo study of bulk hydrogen by treating it as a collection of protons and electrons under the periodic-boundary condition. Find the transition pressure from the molecular solid to simple solid at room temperature. Is it possible to derive the transition pressure of the metallic hydrogen from this simulation?

Bibliography

Anderson M H, Ensher J R, Matthews M R, Wieman C E and Cornell E A 1995 Observation of Bose–Einstein condensation in a dilute atomic vapor *Science* **269** 198–201

Aziz R A, Janzen A R and Moldover M R 1995 *Ab initio* calculations for helium: a standard for transport property measurements *Phys. Rev. Lett.* **74** 1586–9

Bardeen J, Cooper L N and Schrieffer J R 1957 Theory of superconductivity *Review* **108** 1175–204

Bethe H 1931 Zur theorie der metalle. eigenwerte und eigenfunktionen der linearen atomkette *Zeitschrift für Physik* **71** 205–27

Bose S N 1924 Planck's gesetz und lichtquantenhypothese *Zeitschrift für Physik* **26** 178–81

Bradley C C, Sackett C A, Tollett J J and Hulet R G 1995 Evidence of Bose–Einstein condensation in an atomic gas with attractive interactions *Phys. Rev. Lett.* **75** 1687–90

Cramer C J 2002 *Essentials of Computational Chemistry* (New York: Wiley) pp 191–232

Davies K B, Mewes M-O, Andrews M R, van Druten N J, Durfee D S, Kurn D M and Ketterle W 1995 Bose–Einstein condensation in a gas of sodium atoms *Phys. Rev. Lett.* **75** 3969–72

Economou E N 2006 *Green's Functions in Quantum Physics* 3rd ed (Berlin: Springer)

Einstein A 1925 Quantentheorie des einatomigen idealen gases *Sitzungsberichte der Preussischen Akademie der Wissenschaften* **1** 3

Feynman R P and Hibbs A R 1965 *Quantum Mechanics and Path Integrals* (New York: McGraw-Hill)

Hohenberg P and Kohn W 1964 Inhomogeneous electron gas *Phys. Rev.* **136** B864–71

Joly F, Lhuillier C and Brami B 1992 The helium-graphite interaction *Surf. Sci.* **264** 419–22

Kohn W and Sham L 1965 Self-consistent equations including exchange and correlation effects *Phys. Rev.* **140** A1133–8

Laughlin R B 1983 Anomalous quantum hall effect: an incompressible quantum fluid with fractionally charged excitations *Phys. Rev. Lett.* **50** 1395–8

Ma H and Pang T 2004 Condensate-profile asymmetry of a boson mixture in a disk-shaped harmonic trap *Phys. Rev.* A **70** 063606

Ma H and Pang T 2006 Path-integral quantum Monte Carlo study of a mixture of Bose–Einstein condensates *Phys. Lett.* A **315** 92–6

Mattis D C 1993 *The Many-Body Problem* (Singapore: World Scientific)

Pang T 2014 Diffusion Monte Carlo: a powerful tool for studying quantum many-body systems *Am. J. Phys.* **82** 980–8

Pearson S, Pang T and Chen C 1998 Critical temperature of trapped hard-spheres Bose gases *Phys. Rev.* A **58** 4796–800

Slater J C 1974 *Quantum Theory of Molecules and Solids IV* (New York: McGraw-Hill)

www.ingramcontent.com/pod-product-compliance
Lightning Source LLC
Chambersburg PA
CBHW082112210326
41599CB00033B/6672

* 9 7 8 1 6 8 1 7 4 0 4 5 4 *